Assessing Feasibility with Value-laden Models

Simon Hollnaicher

Assessing Feasibility with Value-laden Models

Discussing the Normativity of
Integrated Assessment Models

 **J.B. METZLER**

Simon Hollnaicher
Otto-von-Guericke-Universität Magdeburg
Magdeburg, Germany

The following work has been submitted as a doctoral dissertation at the Faculty of History, Philosophy and Theology of Bielefeld University. Day of the doctoral defense: 10th of July, 2024.

ISBN 978-3-662-70713-5   ISBN 978-3-662-70714-2   (eBook)
https://doi.org/10.1007/978-3-662-70714-2

This J.B. Metzler imprint is published by the registered company Springer-Verlag GmbH, DE, part of Springer Nature.
The registered company address is: Heidelberger Platz 3, 14197 Berlin, Germany

If disposing of this product, please recycle the paper.

Acknowledgements

It takes a village to keep us going, and this is certainly true for writing a book.

I want to thank my supervisor, Prof. Martin Carrier, who supported and inspired me with his curious and encouraging nature. Thanks also to my second supervisor, Prof. Mathias Frisch, for the helpful discussions that shaped this book in important ways. Dominic Lenzi was an inspiration in finding this research question and was always generous with his thoughts. Gratitude to Prof. Kirsten Meyer for her sincere support over the years and her helpful feedback. Thanks to Louis Kohlmann and Valeska Martin for discussing almost every aspect of this book and for being on this philosophical journey with me for much longer; to Hanna Metzen and Johanna Lohrer for reading and commenting on chapters of this book; and to the many from whom I received valuable feedback over the past years. I am grateful to be part of a great research group, the GRK2073, to which I owe many intriguing discussions, warm camaraderie, and good friendships. Thanks to Ulrike and Joachim for hosting me.

Without my family and friends, none of this would have been possible, nor worthwhile. Thank you for everything.

To Stella, whose love and solid ground I returned to every day during this journey. I feel blessed to have you in my life.

Contents

Abbreviations[1]

AR   *Assessment Reports* of the **IPCC**. The ARs give a comprehensive state of knowledge concerning climate change and our response options. The sixth cycle of ARs began with the release of the Physical Science Book in 2021 and ended with the release Synthesis Report in 2023.

BECCS   *Bioenergy with Carbon Capture and Storage.* Widely used **CDR** in mitigation pathways that relies on capturing CO_2 in biomass and sequestering and storing the CO_2 after using the bioenergy.

CB-IAM   *Cost-Benefit-Integrated-Assessment-Model.* A class of **IAMs** that aim to balance the costs (and benefits) of climate mitigation with the costs (and benefits) arising from climate impacts to determine "optimal" emissions trajectories.

CDR   *Carbon Dioxide Removal.* Umbrella term for various techniques that remove CO_2 permanently from the atmosphere.

CEA   *Cost-Effectiveness-Analysis.* In this context the approach of IAMs to compute least-cost pathways for a given climate goal and set of scenario assumptions.

CPA   *Conditional Probability Account.* Account of feasibility developed by Brennan and Southwood (2007) and Gilabert and Lawford-Smith (2012). It states that feasibility is the conditional

---

[1] Integrated modeling is a field rich in abbreviations. As this book will involve some of them, here is a short overview to which the reader can return.

	probability of an agent bringing something about if she seriously tries.
DICE	*Dynamic Integrated Climate-Economy model.* **CB-IAM** developed by William Nordhaus.
FUND	*Climate Framework for Uncertainty, Negotiation and Distribution.* **CB-IAM** developed by Richard Tol.
GCM	*Global Circulation Models.* Complex computer models for simulating the atmospheric and oceanic processes. **IAMs** typically involve a highly simplified emulation of these models in the form of **MAGICC**.
GEA	*Global Environmental Assessments.* A class of reports on global environmental problems, which synthesize knowledge in order to inform decision-making, often going back to intergovernmental structures.
IAM	*Integrated Assessment Model.* Computer-based model that integrates knowledge from at least two, but often more different systems. This book focuses on **PB-IAMs** applied to the climate problem, which typically at least integrates a representation of the energy, land, industrial, and climate systems.
IAMC	*Integrated Assessment Modeling Consortium.* A joint organization of research institutions for integrated assessment modeling and analysis. It was created in 2007 after the **IPCC** called for an independent body to lead the provision of emissions scenarios to keep the **IPCC** as an independent assessor who does not conduct research.
IIASA	*International Institute for Applied Systems Analysis.* Influential institute and research hub for integrated modeling in Laxenburg, Austria. Home of **MESSAGE**.
IMAGE	*Integrated Model to Assess the Global Environment.* Specific **PB-IAM**. Developed at the PBL Netherlands Environmental Assessment Agency in the Netherlands (cf. Stehfest et al. 2021; Roelfsema et al. 2022).
IPCC	*Intergovernmental Panel on Climate Change.* A UN body that aims to assess the state of the knowledge related to climate change and provide knowledge on climate change to decision-makers.
MAGICC	*Model for the Assessment of Greenhouse Gas Induced Climate Change.* A model of the atmospheric and oceanic interactions

with highly reduced complexity. Used in many assessments and typically relied upon by **IAMs**.

MESSAGE *Model for Energy Supply Strategy Alternatives and their General Environmental Impact.* Specific **PB-IAM**. Developed at the **IIASA** (cf. Krey et al. 2020).

PAGE *Policy Analysis of the Greenhouse Effect.* Specific **CB-IAM** developed by Nicholas Stern and used in the Stern Review.

PB-IAM *Process-based Integrated Assessment Model.* Also known as "policy evaluation models" or "dynamic process IAMs". One category of **IAMs**, which is typically more detailed in its representation of the individual systems and focuses on mitigation efforts. PB-IAMs create mitigation pathways compatible with a given temperature goal.

PIK *Potsdam Institute for Climate Impact Research.* Leading hub for **IAM** research and developer of the **ReMIND** model. Based in Potsdam, Germany.

RCP *Representative Concentration Pathways (RCPs)* are "internally consistent sets of projections of the components of radiative forcing" (van Vuuren et al. 2011, 7) developed by the **IAM** community in order to provide a common basis for the different working groups of the **IPCC**. The RCPs are named after the expected forcing at the end of the century. There are five pathways: RCP8.5, RCP6, RCP4.5, RCP2.6 (~ 2 °C warming), and RCP1.9 (~ 1.5 °C warming).

ReMIND *Regional Model of Investment and Development.* Specific **PB-IAM** (cf. Luderer et al. 2020). It is developed by the **PIK**.

RPA *Restricted Possibility Account.* Conception of feasibility developed by Wiens (2015). According to it, an outcome is feasible if it is possible to bring it about, given the all-purpose resources at our disposal and the processes we can use to transform them over time.

SCC *Social Costs of Carbon.* Monetary evaluation of the implication by a single ton of CO_2. A main output of calculations with **CB-IAMs.**

SSP *Shared Socio-Economic Pathways*: Five overarching narratives and background conditions describing plausible alternative developments on key divers such as population and education levels (cf. B. C. O'Neill et al. 2014; Riahi et al. 2017). Serve as an input to **PB-IAMs**.

SWF   *Social Welfare Function.* Main target function of the algorithm in
             many **PB-IAMs** that determines the level of social welfare.
UNFCCC   *United Nations Framework Convention on Climate Change.* UN
             treaty that aims to combat climate change.

Introduction

Questions of feasibility are frequent in our practical thinking. We wonder whether we can reach the train station in time, heal a broken relationship, pay back some pressing loan, or finish the renovation on which we overextended ourselves. Collectively, we discuss how it is feasible for us to contain the spread of a virus (cf. *Hellewell et al. 2020*), whether the global community can achieve its goal of Zero Hunger (cf. *Blesh et al. 2019*), or if there is still a way to keep global warming below 1.5 °C (cf. *Rogelj, Shindell, et al. 2018*). When asking these questions, we typically have already settled that the goals themselves are desirable. The question is whether and how we can bring them about. This is the feasibility question.

We often disagree passionately on answers to these questions. This already indicates that something about feasibility makes it a contested issue. Discussions on these questions may further reveal that we are often not only unsure about a particular answer but also about the meaning of the question itself. Our discussions might slip seamlessly from claims on the achievability of a goal to whether we should pursue it or whether it is realistic to expect it to happen. We wonder what assumptions we are allowed to make about the world and ourselves when making feasibility claims. Expert advice on the feasibility issue is precious, but the confusion surrounding the question and concept suggests that there might also be a role for philosophy to play. Philosophy has always been more interested in clarifying questions than giving a particular answer. It may contribute to contested issues by providing conceptual clarification and reflecting on the practices and methods involved in dealing with these concepts. This is the goal of this book concerning scientific assessments of feasibility.

© The Author(s) 2025
S. Hollnaicher, *Assessing Feasibility with Value-laden Models*,
https://doi.org/10.1007/978-3-662-70714-2_1

This book addresses the *value problem in assessing feasibility*. The question is how we can assess the feasibility of climate goals and mitigation strategies with value-laden models that intrinsically depend on normative assumptions. The models in question are *Integrated Assessment Models*, or IAMs. I will introduce them in more detail below. For now, it suffices to know that IAMs combine knowledge from different fields to give an overarching perspective on climate mitigation and generate comprehensive pathways for the future. IAMs are an influential tool in providing advice on Global policymaking, and scenarios from IAMs have shaped how we think about mitigating climate change in many ways. Chap. 6 will show that IAMs depend on many normative assumptions deeply embedded in the models. This raises the question of how IAMs can provide scientific assessments on feasibility, which many think of as referring to the empirical facts of a situation. I will argue that the concept of feasibility itself contains a value dimension and that modelers must deal with the normativity of modeling feasible futures. However, despite value-laden models, objective and legitimate scientific advice is possible by making normative assumptions explicit, increasing the plurality of value perspectives in such assessments, and deliberating on value aspects with the public. But as normative assumptions within the current scenario evidence are often implicit and cover only a particular corner of the viable value spectrum, the legitimacy of assessments and policy advice with IAMs is compromised, or so I will argue.

Feasibility is a central concept in our practical and political thinking. Understanding more clearly how we can come to feasibility judgments on complex social and global goals seems to be a worthwhile and important inquiry. The case of assessing the feasibility of climate goals with IAMs poses a particularly interesting case for two more reasons.

First, feasibility assessments in the climate case are of paramount *practical relevance*. Our cumulative emissions from burning fossil fuels and land changes have brought the delicate climate system to the brink of collapse. Quitting is without a viable alternative, but even the most radical climate activists will agree that we cannot go cold turkey on carbon emissions. Fossil fuels run through our lives in various ways, and they have fueled the world, for better or worse, over the past 200 years. We need *guidance* on a transition that needs to happen fast and touches on various technological, economic, environmental, social, institutional, and ethical issues. Mitigating climate change is complex, but the basics are understood enough to expect such guidance. The assessments we come up with will likely be consequential. Despite being a relatively small scientific field, integrated modeling has already been shown to greatly influence policymaking and the framing of our response options to the climate problem (cf. *S. Beck and*

Mahony 2018b; Haikola, Hansson, and Fridahl 2019; McLaren and Markusson 2020). We better have a good understanding on what the model tells us and how they come to their assessments of feasible futures.

Second, the feasibility question in the field of climate mitigation provides us with a *concrete* and *tangible* case of assessing feasibility from which we might draw more general insights. The climate case is special, as there are established, concrete goals. In 2015, the world community adopted the Paris Goals, committing to holding global warming "to well below 2 °C" and to "pursue efforts to limit the temperature increase to 1.5 °C" (*United Nations General Assembly 2015*, Article 2). These goals are the moral point of reference for many agents in the field of climate change, activists and leaders alike. What is contested about these goals is not their moral desirability but what is needed to achieve them. It is the feasibility of these goals that is often doubted, for instance, when Guillemot (*2017*) calls the 1.5 °C goal "the necessary and inaccessible" objective. The climate case is, moreover, tangible, as there are established scientific venues and methods for addressing the feasibility question. The IPCC reports are especially noteworthy as they are widely accepted to provide the state of the knowledge regarding climate change, including future pathways and their feasibility (cf. *Rogelj, Shindell, et al. 2018; Riahi et al. 2022*). IAMs play a central role in the chapters on transformation pathways for the future. We may use this case of "assessing feasibility" to understand something more general.

It will be helpful to the reader if I explain certain key concepts upfront. As mentioned, this book investigates IAMs, or *Integrated Assessment Models*. IAMs are "Integrated," as they combine representations of different systems, such as the energy, land, industry, social, and climate systems, to provide an overarching view of climate mitigation. IAMs provide "Assessments," as they are targeted at being relevant to policymaking. As "Models," IAMs abstract and idealize the underlying system of reality in many ways, implemented in computer code. So-called Process-based-IAMs (PB-IAMs)[1] produce *mitigation pathways*, which represent trajectories for the future compatible with a given temperature goal and based on a range of scenario assumptions. I will refer to this mode of modeling as "assessing feasibility." An outcome is *feasible*, roughly speaking, when there is a way we can bring this outcome about. To show that something is feasible, we must provide a viable path from the status quo to it. Such paths can be complex and demanding, but if there is a sequence of actions and events that can get us

---

[1] PB-IAMs *take* a temperature goal as an input, while the other kind of IAMs, Cost-Benefit IAMs, *determine* an optimal climate goal. The book is concerned with the former. I will explain the difference below and in greater detail in Chap. 4.

there, the outcome is feasible. Finally, I will repeatedly describe the models, or science in general, as containing *value judgments*. This is a common term in the philosophy of science, though sometimes the meaning is not entirely made clear. For now, we can think of value judgments as any judgment that depends on some variable or term whose' determination demands a normative justification.

There is an interesting common ground between publications from philosophy, the integrated assessment literature, and some (though not all) of the social science publications on the feasibility issue. I will refer to it as *Descriptive Feasibility* (cf. Chapter 3). It states that feasibility is a descriptive concept, the meaning of which does not depend on value judgments in any way. To determine whether a goal is feasible, we should only look at the empirical evidence. Ethical considerations have no role to play in such assessments. In fact, bringing value judgments to assessing whether a goal is feasible would amount to a category mistake, violating a fundamental distinction between the descriptive and the normative. This might be best made plausible by the example given by Brennan and Southwood (*2007, 8*): "If we lack relevant medical knowledge and expertise, it may not be feasible that we perform a delicate neurological operation on a patient's right hemisphere—even if this is the only way to save her life." There are simply two separate questions concerning the operation: whether it is desirable and whether it is feasible. The goal's desirability makes it no more feasible (and vice versa). Mixing the two in any way makes us prone to confusion.

Many conceptual and substantive contributions on feasibility agree on *Descriptive Feasibility*. Wiens writes, for instance, that "feasibility assessments do not incorporate our judgments about which states of affairs are worth realizing from a moral standpoint" (*Wiens 2015, 9–10*). Other notable philosophical contributions to feasibility also subscribe to its descriptive character (cf. *Brennan and Southwood 2007*; *Cohen 2009*; *Gilabert and Lawford-Smith 2012*). Political philosopher Simmons writes that "[d]eterminations of 'political possibility' and 'likely effectiveness,' [...] seem more naturally to require the expertise of, e.g., political scientists, economists, and psychologists" (*Simmons 2010, 19*) than that of moral philosophers. Contributions from the modeling community and social scientists often also restate this conceptual distinction. Brutschin et al. "stress the importance of a conceptual and operational distinction between feasibility and desirability" (*Brutschin et al. 2021, 2*). Gambhir et al. (*2017*), in their exploration of "the critical notion of how feasible it is to achieve long-term mitigation goals," emphasize that they exclude "political and social concerns" as they would touch on values (*Gambhir et al. 2017, 2*). Distinguishing feasibility from desirability is a core commitment of all sides.

However, the conceptual fixation of feasibility being "void of any moral content" (*Wiens 2015, 9*) stands in a strange tension with philosophical discussions on values in science. Most philosophers of science have accepted that coming to scientific facts depends in various ways on making value judgments. Value judgments are accepted as a legitimate part of scientific assessments. The question is, rather, how to draw the line between those value influences that are legitimate and those that are not (*Holman and Wilholt 2022*). Moreover, in the case of IAMs, in particular, the entanglement of facts with values is hard to deny on closer inspection. Matthew Adler calls integrated modeling an "exercise in ethics" (*Adler and Treich 2015, 279*), and philosophers have pointed out the various ways values appear in integrated modeling (yet this discussion often targets Cost-Benefit-IAMs instead of PB-IAMs, between which there are substantive differences) (cf. *Gardiner 2011; Schienke et al. 2011; Frisch 2013; Kowarsch 2016; Frank 2019; Mintz-Woo 2021b*). The value-dependence of integrated modeling is even becoming recognized by voices within the modeling community. Massimo Tavoni, one of the most cited authors of the IAM community, and fellow modeler Giovanni Valente (*2022*) recently argued that investigating the ethical aspects of the models deserves far greater attention. They write:

> "We posit that the normative components of models—more than the physical and socio-techno-economic ones- are the most fraught by uncertainty and yet the least understood. We suggest a research agenda to explore uncertainties of evaluation frameworks, transcending the current implicit normativity of IAMs" (*Tavoni and Valente 2022, 321*).

One particular instance recently brought ethical questions in mitigation pathways to the forefront of discussions. Pathways for low-temperature goals in the Special Report 2018 relied extensively on negative emissions across the 21st century. Negative emissions ease the transition and promise to make formerly unsolvable goals possible in the models. However, their use in IAMs led to fierce objections, pointing out that such a bet on negative emissions is ethically contentious and that providing these negative emissions will involve problematic side effects (cf. *Peters 2016; van Vuuren et al. 2017; Lenzi et al. 2018*). In a contribution from climate ethics, Shue (*2017*) argues that shifting risk to future generations to relieve the Global North from too challenging near-term mitigation has "unacceptable moral costs" and is akin to a game of Russian Roulette, in which the winner and losers are different groups, a gamble to take or to offer would be "outrageous" (*Shue 2017, 208*). Whether one agrees with this ethical analysis or not, feasibility assessments from IAMs suddenly found themselves in heated justice debates.

Given the value-ladenness of IAMs and science in general, one may doubt that science can deliver feasibility facts at all in light of *Descriptive Feasibility*. I will argue that the relation between feasibility and values is more complex than the conceptual distinction of *Descriptive Feasibility* confers. The question is how scientists can provide legitimate assessments and advice on feasibility in light of models and methodologies that depend inescapably on normative assumptions. This question cannot be concluded on conceptual grounds alone but demands to be attuned to insights from the philosophy of science, or so I will argue.

In doing so, the book engaged with different strands of literature, to which it hopes to provide some insights. *First*, it is a critical engagement with IAMs and, thus, contributes to the interdisciplinary discourse on the models. Contributions from integrated modeling structures around a range of key projects and reports, most notably Chap. 3 of the latest Assessment Report of the IPCC (*Riahi et al. 2022*) and Chap. 2 of the Special Report on the 1.5 °C goal (*Rogelj, Shindell, et al. 2018*). Both chapters contributed to the feasibility question regarding climate goals and mitigation strategies. Feasibility assessments based on IAM scenarios have been developed and brought forward by modelers and social scientists (cf. *Riahi et al. 2015*; *van Sluisveld et al. 2015*; *Gambhir et al. 2017*; *Nielsen et al. 2020*; *Jewell and Cherp 2020*; *Brutschin et al. 2021*; *van de Ven et al. 2023*). Moreover, there is an extensive and growing interdisciplinary debate concerning IAMs (cf. *Pindyck 2013*; *Fuss et al. 2014*; *Edenhofer and Kowarsch 2015*; *Geden 2015*; *Rosen and Guenther 2015*; *K. Anderson and Peters 2016*; *M. Beck and Krueger 2016*; *S. Beck and Mahony 2018b*; *Lenzi et al. 2018*; *Haikola, Hansson, and Fridahl 2019*; *Low and Schäfer 2020*; *van Beek et al. 2020, 2022*; *Keen 2021*). At least some of these discussions touches on what I called the "value problem." There are also important philosophical contributions to the issue (cf. *Gardiner 2011*; *Schienke et al. 2011*; *Frisch 2013, 2018*; *Kowarsch 2016*; *Weyant 2017*; *Frank 2019*; *Mintz-Woo 2021b*; *Lenzi 2021*; *Lenzi and Kowarsch 2021*). It engages critically with the value judgments involved in IAMs, hoping to contribute to the understanding of IAMs and their results. Moreover, it reflects on what principles should guide the modeling community in dealing with value aspects. In this regard, this book contributes to the interdisciplinary discourse on IAMs and climate policy analysis.

A *second* strand of literature of this book concerns the concept and role of feasibility. Political philosophy has become increasingly concerned that focusing on ideal theories makes political theory irrelevant and not helpful in reducing the injustices of the world as it is (cf. *O. O'Neill 1987*; *Mills 2005*; *Simmons 2010*; *Ypi 2010*; *Valentini 2012*; *D. Miller 2013*; *D. Estlund 2020*). Unsurprisingly, this is also of concern in climate ethics (cf. *C. Heyward and Ödalen 2016*; *Caney 2016*). One way to understand this distinction is that non-ideal theories should take feasibility facts into account (cf. *Valentini 2012*), which has sparked a conceptual debate on feasibility itself (cf. *Majone 1975*; *Räikkä 1998*; *Brennan and Southwood 2007*; *Gilabert and Lawford-Smith 2012*; *Lawford-Smith 2013*; *Wiens 2015*; *Hamlin 2017*; *Southwood 2018, 2022*; *Erman and Möller 2020*; *Stemplowska 2020*). This book relies on this debate for conceptual guidance for scientific advice on feasibility. It also aims to take a position by arguing for a thick understanding of feasibility and showing that the kind of feasibility facts that can be expected by scientific assessments (at least in one particular field) will involve value judgments. Non-ideal climate ethics should, thus, not only *take* scientific facts into account in their ethical reasoning but should *engage* with value questions in scientific assessments on feasibility. Critical examination of IAMs and policy analysis is highly important, and philosophers bring the tools and knowledge to foster value transparency and plurality in this highly policy-relevant field.

A *third* thread running through this book is the debate on values in science, mainly discussed in literature from the philosophy of science (cf. *Rudner 1953*; *Levi 1960*; *Longino 1990*; *Lacey 1999*; *Douglas 2000, 2009*; *Putnam 2002*; *E. Anderson 2004*; *Wilholt 2009*; *Elliott 2011*; *Betz 2013*; *Carrier 2011, 2013, 2019, 2021*). As the debate on whether the "value-free ideal" of science can be upheld is coming to an end, the "new demarcation problem" concerns where and how to draw the line between legitimate and illegitimate value influence in science (*Holman and Wilholt 2022*). This book engages with a case of value judgments in science. The case of IAMs is particular, as IAMs aim directly at the policy discourse and rely on a scenario approach. These factors will play a role when discussing IAMs as a case of values in science, and they demand careful distinctions at times. One contribution to this strand is the explication of concrete value judgments in IAMs. Further, the book proposes and defends three guiding principles for dealing with values in scientific assessments and policy advice.

Finally, I hope this book provides valuable insights for the wider public. Mitigation pathways from IAM are central tools used to inform on possible solutions to climate change. Since policymaking and public debate should be informed by scientific knowledge in matters like climate change, we need a clear view of what the model tells us. This book is a small contribution to this task.

The book consists of three parts. *Part I* (Chaps. 2 & 3) discusses the concept of feasibility and explicates a conception that is useful for guiding feasibility assessments on complex social and global goals. It defends a particular explication of feasibility as a thick concept. *Part II* (Chaps. 4 & 5) engaged with how the models assess feasibility. It describes the models, retells the history of how they came to focus on feasibility questions, and critically engages with the methodology of assessing feasibility. The upshot of this part is that modelers need to respond to the value dependence of IAMs. *Part III* (Chaps. 6–8) lays out the value-ladenness of IAMs and explicates various value questions arising in integrated modeling. It discusses how value dependence threatens the legitimacy of scientific advice and argues that by fostering value transparency, plurality, and democratic endorsement, scientists can provide legitimate assessments despite value-laden models. Chap. 8 discusses some implications of these principles for integrated modeling. It argues that the current evidence from IAMs, as applied in feasibility assessments, involves biases and risks perpetuating injustices if unaddressed. Modelers should provide more value-explicit scenarios in order to put the principles into practice. In the following, I will go through the chapters in more detail, providing an overview of the book's argument and key claims.

Chapter 2 argues that we should understand feasibility as a form of restricted possibility as proposed by Wiens (*2015*) and that this conception is a good guide for "assessing feasibility" in the climate context. The chapter starts with four anecdotes of feasibility judgments made by different kinds of agents in the climate policy debate. It goes on to give an initial characterization of feasibility. Feasibility is about how we can bring a particular outcome about. To call an outcome feasible means that there is a trajectory from the status quo to the state of affairs in question, which pays simultaneous attention to the various limitations we face. In the literature, feasibility is taken (a) to rule out proposals and (b) to provide comparative guidance. Two philosophical definitions of feasibility stand out in the literature. The Conditional Probability Account understands feasibility as a scalar concept in terms of the probability of success conditional upon the agent in question trying. I will argue that this conception provides a too narrow framing of how we should understand feasibility assessments of complex social goals. The second definition is Wiens' conception of feasibility as a restricted form of possibility. Wiens understands feasibility as a form of attainability in light of the resources and processes that are available to us. This makes better sense of the pathway character of feasibility and is a helpful explication of the concept for scientific feasibility assessments. However, two open questions remain to be answered by such an account. *First*, how to distinguish feasibility from the merely practically possible, and *second*, how feasibility can provide comparative guidance.

Chapter 3 argues for a value dimension of feasibility and that this dimension helps address the two concerns. The chapter starts by introducing the position of "descriptive feasibility." Viewing feasibility as a descriptive concept promises to provide the ground for separating scientific facts from ethical and political values. While many authors hold this view, the chapter surveys some conceptual contributions that argue for a value dimension to feasibility. It then provides an argument to understand feasibility as value-laden or "thick." I propose a thick conception that builds on Wiens' account. Value judgments are part of feasibility judgments in three ways: value judgments exclude unacceptable means, exclude measures with unacceptable side effects, and are relied upon to define a threshold concerning the acceptable level of uncertainty. Allowing for a value dimension answers the first conceptual challenge of distinguishing feasibility from the merely practically possible. The chapter then concretizes the normative role of feasibility. Feasibility licenses inferences concerning what options are worthy of deliberation. Finally, the supposed comparative sense of feasibility, in which an option is more feasible than other options, is recovered through related value-laden concepts, such as some options being "more challenging" or "more realistic," meeting the second challenge.

Chapter 4 introduces the class of models. It provides a short history of Integrated Assessment Modeling, which goes back to computer-based assessments in the 1970 s. Two classes of IAMs must be kept distinct. Cost-Benefit-IAMs (CB-IAMs) generate pathways that aim for an optimal balance between costs of mitigation and climate impacts and have been subject to extensive philosophical debate. The second class of models is *Process-based-IAMs* (PB-IAMs), which are an influential part of the science-policy interface but have yet to see much engagement from climate ethics. The main difference between the models is that PB-IAMs take climate goals as a fixed input. PB-IAMs model so-called "mitigation pathways" to reach these predefined temperature goals. The chapter describes PB-IAMs in more detail and provides the "case for IAMs." IAMs are valuable as they provide unique access to model the interactions and dynamics at the core of mitigating climate change. The chapter explains how IAMs rely on scenarios and ends by describing the most recent part of the history of PB-IAMs, in which they are increasingly used to assess the feasibility of climate goals and different pathways.

Chapter 5 addresses the question of how IAMs can assess feasibility. It provides two reasons, rooted in the concept of feasibility as explicated above, which put the models in a good spot to provide feasibility claims. Making judgments about the feasibility or infeasibility of a complex social or global outcome is special as it *(a)* demands to have a perspective that takes all relevant constraints into account simultaneously and *(b)* must allow for complex pathways in which we can change many aspects of our environment over time. As the strength of IAMs lies in these two regards, they are promising (though clearly imperfect) scientific tools for assessing the feasibility of climate goals. Solvable scenarios in IAMs can be understood as evidence for the feasibility of these goals. Evidential relations depend on background assumptions. This reflects the contextual aspects of feasibility judgments, which are often made in view of certain implicit beliefs. For a scientific assessment of feasibility, we need reliable methods to evaluate these background assumptions. The chapter will argue that the existing practices of model evaluation in relation to IAMs cannot provide empirical support for or against these background beliefs. As they are used for empirical testing, so-called "appeals to the past" fail to provide the right kind of knowledge, and empirical estimations of constraints involve unreasonable high uncertainty. The chapter discusses and rejects one recent empirical framework of assessing feasibility with IAMs in Brutschin et al. (*2021*). The chapter argues that feasibility assessment must pay closer attention to value judgments, bringing in the perspective of values in science to provide a more solid ground for assessing feasibility with IAMs.

Chapter 6 provides a taxonomy of value judgments in IAMs. It discusses eight aspects of integrated modeling through which value judgments become embedded in scenario evidence from IAMs. The chapter goes through the following aspects of modeling: *(1)* Value judgments occur when feasibility indicators and constraints are chosen, either in assessment frameworks or within the models themselves. *(2)* Agenda-setting involves value questions concerning what kind of evidence is produced. *(3)* PB-IAMs generally rely on cost-effectiveness as a framework for modeling pathways, which is only one possible criterion for sharing the burden of climate mitigation. *(4)* An influential value judgment concerns the conceptualization of well-being in the models, which determines how costs and burdens are represented in the models (and which ones). *(5)* Related, IAMs use mechanisms that touch on aspects of inequality and fairness between regions and within a region. *(6)* Discounting presents a value judgment on evaluating costs and burdens occurring over time. *(7)* IAMs must define a modeling domain, which presents another value judgment. *(8)* Finally, uncertainty in modeling parameters and assumptions gives rise to inductive risk. I analyze a case of this in the reliance on Carbon Dioxide Removal techniques. For each of these eight aspects, I explicate the value judgments in integrated modeling and, as best as possible, put the most common practices into context and contrast them with alternative value outlooks.

Chapter 7 discusses what the goal of objectivity of scientific assessment implies in the face of value-laden models. The chapter starts by explaining that scientific policy assessments face a problem of legitimacy if they are implicitly bound up with specific value outlooks. Scientists then risk having an undue influence on policy decisions. To prevent this, scientists must seek objective and neutral advice despite value-laden models and methods. The chapter proposes three guiding principles for legitimate scientific assessments with IAMs. The first principle is that value judgments must be made *transparent*. Ideally, influential value judgments must be presented as explicit premises of the results. The second principle is that integrated assessments must increase the *plurality* of value outlooks represented in IAM scenarios. All viable value outlooks concerning different mitigation strategies should be backed up by some piece of scientific evidence. The third principle is that modelers should seek public engagement on value aspects of the models and their scenarios. Deliberative venues can legitimize value judgments in scenarios by providing a form of *democratic endorsement*. These three principles provide a provisional solution to the problem of legitimacy.

Chapter 8 concludes the discussion of this book. It criticizes the current state of evidence from IAMs as biased and at risk of perpetuating existing injustices under the mask of neutral and objective feasibility assessments. The chapter described three biases that mark the tendency of the models to shift burdens to future generations, away from current major emitters, and to favor entrenched interests. These biases risk perpetuating injustices, as IAM pathways are powerful representations of the future. The models' internal tendencies are masked under the veil of feasibility in two ways. Value-laden assumptions are often made to make pathways "more realistic" and thus disguise themselves as feasibility judgments. Moreover, presenting IAM results as "feasibility assessments" conveys an image of descriptive and neutral scientific assessment, which masks its value-dependence. The chapter provides a range of concrete implications of these discussions, most notably the need to produce ethically explicit scenarios. Justice questions must be part of the scenario design, and ethicists should engage with them more actively. Finally, I recall the self-acclaimed metaphor of modelers as mapmakers, arguing that the maps to envisage (if one relies on this metaphor) must be maps where the value-dependent conventions and standards of the mapping are an explicit part of the map.

Part I
On the Concept of Feasibility

Feasibility

Before delving into discussions on the meaning and role of feasibility, let me set the stage by introducing four short anecdotes, each describing an instance in which feasibility claims played a critical role in the recent climate discourse.

2.1 Four Examples of Feasibility Claims

BERLIN2030

In 2022, the civil movement "Klimaneustart" was trying to initiate a referendum in Berlin on whether the city should tighten its climate target to being climate neutral by 2030 instead of 2045. The legal procedure demands that activists first provide a certain amount of signatures as a sign of public support. By October of that year, the movement had collected about half of the 175.000 signatures needed to force the referendum with about six weeks left. Public interest in the referendum increased, and a challenging question for citizens and policymakers became whether the stricter climate target was feasible. In this context, the Green Party hosted a public discussion on the issue and invited Bernd Hirschl, the lead author of a feasibility assessment that laid the ground for the existing climate goal of reaching neutrality by 2045 (*Hirschl et al. 2021*). Hirschl pointed out various challenges that cannot be met, in his opinion. For instance, the rate of energy-efficient retrofitting of buildings would need to multiply from below 2 % to 10 % per year despite a shortage of material and skilled labor that can hardly be overcome. Due to constraints such as this, he argued, turning Berlin

© The Author(s) 2025
S. Hollnaicher, *Assessing Feasibility with Value-laden Models*,
https://doi.org/10.1007/978-3-662-70714-2_2

climate-neutral within eight years would be infeasible. Such claims of infeasibility loomed large over the movement. The newspaper Tagesspiegel headlined days before the referendum: "scientist judges climate neutral Berlin by 2030 to be infeasible,"[1] eluding to an interview with sociologist Fritz Reusswig. The activists tried to counter such judgments by pointing to cities worldwide with similar targets and referring to other events that appeared infeasible at the time but became a reality, such as the NASA program of landing on the moon in the 1960 s. The activist also tried to change the framing, arguing for the moral necessity of the more ambitious goal. Nevertheless, feasibility remained a primary concern for the campaign. Ultimately, the referendum was unsuccessful, as it failed to gain enough positive votes once it took place.

CLIMATE CITIZEN COUNCIL

Citizen councils often receive less attention than they deserve. A newspaper article published in the ZEIT on the 26th of June 2021 gave a close-up report of one such initiative, "Bürgerrat Klima" ("Citizen Council Climate"). The Bürgerrat Klima is a deliberative body consisting of randomly chosen citizens who debate on issues of climate policy and provide recommendations to the government. Protest movements often demand such citizen councils, and they are slowly becoming a recognized part of climate governance. Recommendations by citizen councils, while typically not binding, carry some democratic legitimacy with them due to being perceived as representative of the general public. Multiple deliberation sessions allow participants to exchange reasons and this promises to help overcome entrenched positions. An essential part of such processes is the involvement of scientific experts. The article describes a short part of the discussion (my translation):

> "In the plenary session, in which all 40 members of this 'field of action' have gathered, the moderator made clear: Only with a reduction in livestock of 80 % can the 1.5 °C goal be reached. But that would mean: no more meat. 45 % reduction would mean: moderate meat consumption, 1.7 °C. A survey of the general mood is obtained, hand signs. Result: only half would be ready to stick to the 1.5 °C goal under these conditions." (*Theile 24.06.2021*)

The moderator here translates findings presented by experts, who, likely based on scenario pathways, deduced the necessary mitigation steps towards staying within 1.5 °C. The feasibility claim concerns the relationship between concrete

---

[1] "Wissenschaftler hält klimaneutrales Berlin bis 2030 für nicht machbar", Tagesspiegel (*23.03.2023*). Reusswig nevertheless supported the initiative.

measures and the goal's attainability: staying within the 1.5 °C pathways is not feasible if one cannot reach an 80 % reduction in livestock.

GERMAN CONSTITUTIONAL COMPLAINT 2021

Court cases concerning national climate policies have significantly increased in recent years, and such juridical evaluations of climate policies must often use scientific knowledge in one way or another. One example is the landmark ruling by the German Constitutional Court in 2021, which criticized the national policies by pointing out that they potentially violate future generations' rights to freedom. The court argued that future generations must be safeguarded from highly restrictive mitigation measures becoming necessary because such measures could seriously undermine the ability of future people to determine their own lives. Since the government was not making its plans explicit beyond 2030, the current climate policy risks using up all of the carbon budget remaining for Germany's share of the Paris Commitment. While the court stayed short of rebuking the climate policy directly, they demanded that the chosen budget and pathway be made explicit. The ruling explicitly cites the IPCC report's calculations and the assessment by the Sachverständigenrat für Umweltfragen, a leading scientific advisory body (*SRU 2020*). The SRU determined a national carbon budget based on a per capita distribution of the remaining Paris-compatible budget (*SRU 2020, 52*), and the court relied upon them over several pages in their reasoning. As the government stays silent on its assumptions concerning the carbon budget, the court asks them to specify their policy plans from 2030 onwards for keeping an emission pathway towards climate neutrality in line with the Paris Goals, which does not risk violating future generations' right to self-determination. In its argumentation, the court thus alluded to a corridor of feasible pathways for Germany compatible with staying within the Paris Goals.

"LÜTZI BLEIBT"

The Paris Goals are also an important reference point for climate activists worldwide. Demonstrators often refer to the 1.5 °C goal, holding governments worldwide accountable to what they agreed upon in 2015. One such example is the German protest against the Garzweiler II coal mine, which culminated in early 2023 in a mass protest in the small village of Lützerath. Lützerath is the latest and last village that falls victim to the lignite mining of Garzweiler II, run by RWE and located in the East of North Rhine-Westphalia. The protest was not purely symbolic. The protesters argued that burning the coal under the area in question was inconsistent with the Paris Goals. Figure 2.1 shows a large banner on one of the main walls in the occupied village, stating, "1.5 °C means: Lützerrath stays!" In its most literal interpretation, this statement claims that if the coal in the area of Lützerath is burned, 1.5 °C becomes infeasible. This question of feasibility gained considerable media attention, especially after Pao-Yu Oei et al. (*2023*) calculated that the German emission budget left for 1.5 °C allows only an additional 49 *MtCO₂* of emissions from Garzweiler II (cf. *Rieve et al. 2021*) and the RWE plans, including the coal below Lützerath, would lead to 280 *MtCO₂* being released, five times the allotted budget. The study finds that RWE's plans are "incompatible" with the German commitment to the Paris Goals, backing up the protester's feasibility claim.

Fig. 2.1 Banner in Lützerath on a squatted farm in 2021. © Superbass / CC-BY-SA-4.0 (via Wikimedia Commons), URL: https://commons.wikimedia.org/wiki/File: 2021–11-29-Luetzerath_Proteste-6265.jpg

These examples, taken from climate discourse of the past years, bring out the various forms and contexts of feasibility claims. Feasibility claims, especially if backed up by scientific experts, play an influential role in the political and legal debate on policy. They are used by activist movements ("LÜTZI BLEIBT") or against them (BERLIN2030), feature prominently in public debates on complex policy issues (CLIMATE CITIZEN COUNCIL), and play essential roles in law cases (GERMAN CONSTITUTIONAL COMPLAINT 2021). The (limited) sample given in this section gives an impression of the form and influence of feasibility claims in policy discourses.

The following section will briefly characterize feasibility and its normative roles (Sect. 2.2). The subsequent section will discuss the most prominent account of feasibility, which understands feasibility as the conditional probability of success. This understanding is too restrictive for guiding feasibility assessments in the climate context (Sect. 2.3). A fitting guide is the account of Wiens (*2015*), which understands feasibility as a restricted form of possibility. This account can ground scientific assessments of feasibility and make sense of the dimensions of feasibility relevant to climate mitigation but leaves two challenges open (Sect. 2.4).

2.2   An Initial Characterization of Feasibility

We encounter feasibility claims all the time. In our everyday life, as well as in political life, we often refer to what is possible or impossible for us to do, what is achievable, realistic, viable, practicable, doable, or what somebody can or cannot do. Often (though probably not always), these terms refer to the *feasibility* of a particular outcome or course of action. This section will give an initial characterization of feasibility and describe the key conceptual commitment concerning feasibility that I subscribe to in this book.

This chapter aims to find a conception of feasibility[2], which helps guide scientific assessment while retaining similarity to its common usages. I want to note that this gives us *one* possible explication of the concept, not an analysis of *the* concept of feasibility.[3] This chapter, as well as the next one, sets out to discuss the concept of feasibility. Some have argued to drop the notion of feasibility in the climate context due to its ambiguity (cf. *Lenzi and Kowarsch 2021*). This book takes a different route. The question of feasibility is a contested issue in climate

---

[2] For some early discussions, cf. Majone (*1974*), Majone (*1975*). For more recent contributions see Räikkä (*1998*); Brennan and Southwood (*2007*), Gilabert and Lawford-Smith (*2012*), Gilabert (*2012*), Lawford-Smith (*2012*), Lawford-Smith (*2013*), Brennan and Sayre-McCord (*2016*), Southwood and Wiens (*2016*), Hamlin (*2017*), Erman and Möller (*2020*), Stemplowska (*2020*), Wiens (*2015*). A good introduction to the "feasibility issue" is provided by Southwood (*2018*).

[3] Carnap described the method of explication and gave four desiderata for such an enterprise, neither being necessary nor sufficient. He states that an explication should be sufficiently *exact* (linguistically clear), *similar* to the non-scientific term, *fruitful* in scientific and philosophical theorizing, and *simple* (*Carnap 1945, 1947, 1950*).

policy discourse and features centrally in IAM publications and IPCC reports. Thus, we need a way to understand such claims more clearly. This chapter aims to find a conceptual guide for feasibility assessments in the realm of climate mitigation, which is sensitive to common usages of the term (as in the anecdotes above) and provides a basis for discussing the methodology of assessing feasibility with mitigation pathways from IAMs. Chap. 3 will extend this discussion by arguing for a value dimension of feasibility.

In most general terms, feasibility refers to the *realizability* of some state of affairs or institutional scheme. It is a modal concept attributed to a state of the world that has yet to be the case. In contrast to a pure form of possibility, however, feasibility is linked to our capabilities as agents who may bring about a certain outcome or goal. It might be possible that a volcano erupts, but it makes no sense to speak of feasibility in this case. The feasibility question only emerges because we, as agents, strive to realize some state that we deem desirable.[4] Such a state is feasible "if there is a way we can bring it about" (*Gilabert and Lawford-Smith 2012, 809*).

Feasibility must be understood in terms of a *trajectory* from "here" to "there," from the unsatisfactory status quo to the better state of affairs in the future. As things are most often complicated, better futures cannot be brought about instantly, but we need to embark on multiple measures in a sequence, often standing in complex interdependencies with each other. This sequence of steps and measures must be feasible for the outcome to be. Feasibility claims are thus about trajectories in some sense.[5]

Beyond being linked to human agency, feasibility is often considered a more substantive modal predicate than the "merely practically possible." Certain things are possible to achieve but are nevertheless not feasible. For instance, I may win the lottery and buy a private plane with the money, but buying a private plane is certainly not feasible for me (cf. *Southwood 2018*). Consider BERLIN2030. If

---

[4] Such states of affairs are often "middle objects" somewhere between concrete actions and ethical principles (*Brennan 2013, 319*). There must be something good about the outcome from the agent's perspective, as the feasibility questions mainly arise because we strive for something that we deem desirable. This outcome need not be good overall. Luckily, for instance, it is infeasible for the average Joe to build an atomic bomb.

[5] One other class of feasibility facts concerns the internal stability of some desirable institutional scheme, for instance, in asking whether socialism is feasible not in the sense that we can bring it about but on whether it can exist over time once it is implemented (*Cohen 2009*). Stability could also be considered a conceptual part of "getting there" (*Gilabert and Lawford-Smith 2012*).

pressed, the expert, who was pessimistic that climate neutrality by 2030 was feasible, might agree that Berlin might succeed in reaching climate neutrality by 2030 if it tried, for instance, in case some lucky inventions make clean fuels widely available or if the local government starts acting in an eco-authoritarian way, or if the constituency suddenly makes climate neutrality their absolute political priority. However, such assumptions appear inappropriate. Thus, the experts might hold that these trajectories make the outcome possible but not feasible. I take it that most usages of feasibility have some distinction to the "merely practically possible." Let me call this problem the question of *defining a threshold*, where possibility turns into feasibility. That there must be such a threshold is a first commitment for a conception of feasibility that I want to stipulate.

A second commitment is the context-dependence of feasibility (cf. *Gilabert and Lawford-Smith 2012*). As they are regularly made, feasibility claims involve assumptions about where we stand and what can be reasonably expected for the foreseeable future. Many such dependencies stay implicit, as they are often taken for granted in the context of utterance. For instance, I might say that it is feasible for us to reach the restaurant in time and implicitly take into account that you cannot walk long distances due to a foot injury (but at the same time assuming that the subway runs according to its usual timetable). Feasibility claims always take some things for granted. Compare the influential linguistic analysis by Angelika Kratzer of the concepts "must" and "can," which takes sentences involving these terms as relative to specific aspects "in view of which" some things can or must be done (*Kratzer 1977*). Something, similarly, is feasible in view of certain implicit assumptions about the world. This is my second commitment: Feasibility claims are relative to a particular context and often involve implicit assumptions concerning this context, in light of which they are intended to hold.

A third commitment is that feasibility is foremost a political concept relevant in politics, policymaking, and other complex social settings. This partly springs from my aim of finding a useful conception for guiding feasibility assessments. Certain aspects are more important once we consider political examples. For instance, complex interdependencies and trade-offs often exist between different means and proposals, which should be part of feasibility assessments. Such aspects do not necessarily appear pressing when considering more tidy, small-scale examples from moral theory.

Before moving to the two main conceptions in the literature, let me quickly talk about the normative role of feasibility. I will return to this role in more detail in the next chapter. Feasibility is a concept that has normative significance. In a first sense, it is normatively significant because it shows us paths toward sought-after states in the future. The feasibility question arises because we are faced with the question of what to do (in the world we inhabit and with the limited power we have).[6] As human agents, individually and collectively, we must decide what to do. These decisions matter as we start in a world that requires improvement and is full of injustices. Justice demands to be realized, and how this can be done is the task of feasibility assessments.

More directly, though, authors prescribe certain normative roles to feasibility judgments. Gilabert and Lawford-Smith (*2012*) provide a distinction of two roles that feasibility claims play. First, feasibility claims *rule out* proposals. This might be the most common way to use feasibility claims, and it is akin (if not equivalent, cf. *Southwood and Wiens 2016*) to the ought-implies-can provision. We are calling something infeasible to rule it out as a political proposal. If something is infeasible, we have no moral duty to bring it about or to consider it further when deciding what to do. This role requires a binary sense of feasibility, according to which something is either feasible or not (cf. *Gilabert and Lawford-Smith 2012*; *Southwood 2022, 136*). Such a ruling-out sense is used in the example "Lützi Bleibt". Activists declare further coal mining incompatible with keeping warming below 1.5 °C, making a claim of binary infeasibility. The expert in the case Berlin2030 also makes a binary claim of feasibility, saying that the movement's goal is infeasible. Implicitly, he thereby advises the government to rule out the proposal. All too commonly, non-experts rule out opposing proposals in political debates, often with very little justification, escaping the more difficult task of engaging with them on normative grounds.

The fact that infeasible proposals are ruled out sets the bar of epistemic justification high. Judging that a proposal is infeasible puts it off the table, and we better have good reasons to do so. The ruling-out role has a second implication. It limits the role considerations about the agent's motivation can play. Infeasibility must be distinct from unwillingness. If we allow to rule out proposals due to the agent's unwillingness, we will excuse her from moral duties she is perfectly capable of doing but just unwilling.[7] Feasibility judgments thus must be justified with reference to the facts of the situation. As Majone puts it, constraints are

---

[6] Of course, there are other important normative questions besides "what to do," for example, what attitude to have or what would be ideally just. Feasibility might be a constraint on justice itself (but cf. *Southwood 2019*). These questions are worthwhile and plausibly fulfill various valuable functions (*Berkey 2021*; *D. Estlund 2020*). Maybe feasibility bears some relevance to them as well. However, I will not be concerned with these questions but will focus on the role of feasibility in the normative analysis of what we ought to do, all things considered. This is the question policymakers face when concerned with the challenge of climate mitigation.

[7] This is a central critique of climate ethicists' engagement with existing infeasibility claims in the literature. Most prominently, Posner and Weisbach (*2010*) argue that no international treaty is feasible which does not advance the interest of the all (including our) nations. This

"feature[s] of the environment that (a) can affect policy results, and (b) [are] not under the control of the policy maker" (*Majone 1974, 261*). To the degree the agent's motivation is under her control, it is not a feasibility constraint.

The second role assigned to feasibility judgments by Gilabert and Lawford-Smith (*2012*) is to compare proposals regarding their relative realizability. Feasibility, they write, "enables comparative assessments of various proposals," and this role "is invaluable when it comes to decisions about pursuing political reform" as it is more fine-grained and flexible than the mere gatekeeping role above (*Gilabert and Lawford-Smith 2012, 812*). In this *comparative role*, feasibility is not only concerned with whether something is feasible strictly speaking but whether it is more feasible than something else. This role is somewhat tailored to the author's own account of feasibility, one that I will argue against below. However, the more general insight of this role is worth keeping. When an outcome is more challenging, faces higher hurdles, or is "less feasible" than something else, we should consider this fact in some way when deciding what to do. Feasibility assessments have some guiding function when deliberating what to do. In the second role, feasibility is normatively significant since investigating it brings forward various considerations about a pathway toward a goal that we need to evaluate when deciding what to do. Feasibility claims often have some action-guiding role and a conception should be able to say something about how this can be achieved.

Summing up, this section characterized feasibility as a term concerned with trajectories from the unsatisfying status quo to some more desirable state of affairs. Something is feasible if we can realize it from where we are. Three commitments of this chapter are that feasibility is more substantive than mere practical possibility, depends on contextual assumptions, and is foremost applicable to the political sphere. Feasibility is an important normative concept, as it guides us in realizing a better state of affairs, as the judgment that a proposal is infeasible rules it out, and as there is some action-guiding sense of feasibility beyond this.

---

does not seem right, at least about ourselves. Feasibility should not be conflated with "unwillingness" (*Caney 2010, 128; Clare. Heyward 2012, 2*). As D. Estlund (*2020*) writes, we will never do a chicken dance in front of our class, but it is totally feasible for us to do so.

2.3   The Conditional Probability Account of Feasibility

The most influential account of feasibility that has emerged is the *Conditional Probability Account* (CPA).[8] The CPA argues that we should understand feasibility in terms of the probability of success of an agent to bring the outcome in question about. I will argue that the CPA is not a good guide for policy advice on the feasibility question in the context of climate mitigation as its understanding of feasibility is too narrow for guiding scientific assessments.

The CPA was first introduced by Brennan and Southwood (*2007*), who proposed that a state of affairs is feasible for an agent (or a set of agents) if it would be "reasonably likely" that they succeed in realizing x if they tried. For instance, it is feasible for a surgeon to perform a heart surgery if she is likely to succeed if she tries. Gilabert and Lawford-Smith (*2012*), develop this idea further. The conception the two authors propose is noteworthy as it is often relied upon in discussions of non-ideal climate justice (cf. *Caney 2016, 18*) and in contributions from the social sciences looking for a conceptual guide concerning feasibility (cf. *Jewell and Cherp 2020; Brutschin et al. 2021*).

Gilabert and Lawford-Smith (*2012*) introduce feasibility as a two-prong concept relating to the two roles introduced above. Feasibility can rule out proposals based on a binary sense, according to which some outcome is either feasible or infeasible for an agent. This is the sense Brennan and Southwood (*2007*) have in mind when they characterize feasible outcomes as ones that have a reasonable chance of being realized by us, given that we seriously try. A distinguishing feature of the CPA is the reliance on conditional probabilities. Simply probabilities will not do, as it is crucial not to rule out proposals because an agent is unwilling. If Adam is a skillful but lazy surgeon, a successful surgery might be unlikely because he will not seriously try. It would nevertheless be feasible for him. Thus, the CPA introduces the "if they try"-clause to get things right here (*Brennan and Southwood 2007, 9*).

An obvious shortcoming of Brennan and Southwood's characterization is that it lacks clear criteria for what counts as a "reasonable chance" of success. Therefore, Gilabert and Lawford-Smith (*2012*) take a slightly different route and introduce the distinction between "hard" and "soft" constraints. Hard constraints are such "that they will always be constraints" (*Gilabert and Lawford-Smith 2012, 813*). Such constraints arise, for instance, from natural laws, logical inconsistencies, and similar unchangeable features of reality. They write: "logical and nomological constraints are obviously appropriate" as hard constraints, but also

---

[8] Cf. also Lawford-Smith (*2013*), Gilabert (*2017*), and, as a variation, Stemplowska (*2016*).

physical constraints, and maybe biological constraints in the sense of some "human nature." Something is infeasible in the binary sense only if the proposal conflicts with hard constraints: "It is feasible for [an agent] X to j to bring about [an outcome] O in [circumstances] Z only if X's j-ing to bring about O in Z is not incompatible with any hard constraint" (*Gilabert and Lawford-Smith 2012, 815*).[9] Only if a proposal is infeasible in this sense is it ruled out, according to the CPA.

Binary feasibility is though "a fairly blunt tool" (*Gilabert and Lawford-Smith 2012, 812*), and thus, what is more important, according to the CPA, is a second sense of feasibility they call "scalar feasibility."[10] Scalar feasibility takes a proposal to be more or less feasible depending on how much it conflicts with soft constraints. According to the authors, this sense fulfills the role of comparing different proposals, as scalar feasibility can provide a *ranking* of alternatives concerning their conditional probability of success. An option is "more feasible" than an alternative if the chances that we realize it are higher in comparison.

Scalar feasibility arises in light of what Gilabert and Lawford-Smith (*2012*) calls "soft constraints." Soft constraints "do not rule out, but [...] make outcomes comparatively less feasible" (*Gilabert and Lawford-Smith 2012, 813*). Such soft constraints include economic, institutional, and cultural constraints, such as budget restrictions or institutional rules. The critical feature of soft constraints is that they can be changed over time, which introduces a dynamic or diachronic element to feasibility: "[N]ot everything that is less feasible now (in the comparative sense) need be as infeasible later" (*Gilabert and Lawford-Smith 2012, 814*).[11] Soft constraints can be seen as fulfilling the "pathway character" of feasibility, as they introduce a sense of trajectories of actions towards an outcome. Ultimately, the meaning of scalar feasibility though comes down to the overall probability of success, according to the CPA: "It is more feasible for [an agent] X to bring about [outcome] O_1 than for Y to bring about O_2 when it is more probable, given

---

[9] In a separate article, Lawford-Smith (*2013*) expresses binary feasibility as there is an action for which the "probability of the outcome given that action is greater than zero" (*Lawford-Smith 2013, 251*). This is akin since an outcome might be understood to have a probability of zero if and only if it violates hard constraints. The two definitions are different ways to spell out the same idea.

[10] Gilabert and Lawford-Smith (*2012*) might thus be understood to reject the substantiveness of binary feasibility claims, my first commitment.

[11] By including them as soft constraints, the authors try to balance two dangers: to fall into "cynical realism" on the one hand by taking such malleable constraints too seriously and to avoid "impotent idealism" by neglecting them altogether. They write: "These dynamic responses neither discount the existence of soft feasibility constraints nor treat them as unyielding limitations to change" (*Gilabert and Lawford-Smith 2012, 815*).

soft constraints, for X to bring about O_1 given that he or she tries, than it is for Y to bring about O_2 given that s*he tries" (*Gilabert and Lawford-Smith 2012, 815*). To sum up, the CPA considers feasibility as the conditional probability of success. It distinguishes binary from scalar feasibility claims depending on the nature of the constraints.

By putting the scalar sense at the heart of their proposal, Gilabert and Lawford-Smith (*2012*) try to square the tension between feasibility's influential normative role and the often epistemic uncertain terrain of making feasibility claims. This is one reason the CPA has been welcomed as a good and helpful conceptual guide (cf. *Schuppert 2021*; *Brutschin et al. 2021*). However, the implications of the CPA appear misleading when applied to questions of feasibility in the context of complex social goals. The CPA account captures simple moralistic cases much better than providing a foundation for assessing feasibility in a large-scale social context.[12] While cases in moral philosophy have clear agents and actions and may seem empirically simple enough to warrant a probabilistic understanding of the available courses of action, this is unachievable in the complex terrain of climate mitigation strategies. Boran and Shockley (*2021*) note this problem of the CPA:

"This approach to feasibility [CPA] might be operationalizable in limited scale decision-making processes, but it does not capture the formidable complexity of climate change. This is particularly pressing when it is recognized that climate change is not a discrete issue, but part of a nexus of interconnected planetary challenges" (*Boran and Shockley 2021, 36–37*).

The CPA suggests an expert-based overall assessment of feasibility, which can later be combined with normative considerations by calculating expected moral value: "Considerations about what is more or less feasible must be balanced against considerations about what is more or less desirable in order to identify the political options that have maximal expected normative value" (*Gilabert and Lawford-Smith 2012, 818*). Aiming for exact probabilities, though, risks masking uncertainty and complexities (*Houston 2021*), a feature that can be arguably witnessed in the feasibility framework by Brutschin et al. (*2021*), which takes the CPA as a guiding conception and which I will discuss at length below.

---

[12] The examples used in discussing feasibility are often cases from moral situations in the sense of clear agents and actions. Compare Brennan and Southwood (*2007*), discussing a difficult neurological operation for a surgeon, or the example of D. M. Estlund (*2008, 266*), where several people need to push a car out of a snowstorm (cf. *Gilabert and Lawford-Smith 2012, 819*).

On a conceptual level, the CPA is not a fitting guide due to three aspects. *First*, feasibility in the CPA is too closely linked to agents. Feasibility in the CPA is taken to be an agentive-modal, that is, a predicate, which refers to the abilities of agents similar to statements such as that somebody *can* swim or *is able to* play the piano (*Mandelkern, Schultheis, and Boylan 2017*). The CPA takes political agents to be a broad category encompassing both "individuals (such as the residents or citizens of a state) or groups (such as social movements, political parties, firms or a state's agencies)" (*Gilabert and Lawford-Smith 2012, 812*). Still, outcomes not attributable to a (potential) agent are not captured well by the CPA. Such cases do routinely count towards the attainability of a complex goal. Sometimes, multiple agents temporarily work together to achieve an outcome without forming a group agent (cf. *Southwood 2018*). At other times, it is precisely the forming of a collective agent that is a feasibility concern, but this concern is not attributable to an agent (cf. *Stemplowska 2016*).

The CPA explicitly focuses on clearer cases of agency. Lawford-Smith writes: "What we want to avoid is having such a permissive account of available actions that outcomes like ending global poverty or achieving global carbon neutrality come out as feasible [...] Ending global poverty and achieving global carbon neutrality are both possible. But we do not want to say these things are feasible" (*Lawford-Smith 2013, 250*). The authors thus propose not to speak of feasibility in this context, but this makes their conception inapplicable for many political projects that we care deeply about (including their own example of abolishing slavery). For the context of climate mitigation research, a central research question of the IPCC, namely if and how certain climate goals and pathways are feasible, would be a linguistic confusion.[13]

Second, besides demanding clear agents, relying on the CPA risks focusing too much on individual acts and outcomes. One of the central questions concerning feasibility in scientific assessments is what Hamlin (*2017*) calls "co-feasibility." Co-feasibility refers to the interdependence of multiple actions in terms of feasibility. A set of actions is "co-infeasible" if not all actions can be undertaken, even though every single action is feasible by itself. For example, this can happen when two actions depend on the same resource. (We could also think of

---

[13] Take the question of the IPCC SR1.5, if it is feasible to limit global warming to 1.5 °C. An answer to this question cannot take the form of an agentive-modal since there is no agent to which we ascribe that outcome. To limit the feasibility question to the likelihood of success of an international binding agreement formed by the Conference of the Parties (COP) would be way too narrow. One might consider the CPA as providing some input to feasibility assessments of complex outcomes.

the opposite when each of two actions is only feasible if both are performed.)[14] The account given by Gilabert and Lawford-Smith (*2012*) predicates feasibility to individual acts and outcomes and thus guides us toward assessing feasibility in an isolated fashion. Hamlin points out that such a focus on "act-by-act feasibility will tend to miss issues of co-feasibility, which are nevertheless genuine issues of feasibility" (*Hamlin 2017, 6*).

Third, if we follow the CPA in only allowing clear-cut cases of hard constraints such as logical constraints, we are left with an "ultra-thin" (*Southwood and Wiens 2016, 3048*) sense of binary feasibility that seems not to capture normal use of feasibility at all (for instance, in the examples at the beginning of the chapter). When experts and laypeople call something infeasible, they often mean something more complex than logical, physical, or biological impossibility. Brennan and Southwood take this distinction into account: "Many actions are logically possible yet infeasible. There is nothing logically or nomologically impossible about a medical ignoramus successfully performing a neurological operation for which, as it so happens, he lacks the relevant expertise" (*Brennan and Southwood 2007, 9*). Similarly, Southwood (*2018*) notes that it is clearly infeasible "for me to single-handedly solve the Israel - Palestine conflict, eradicate poverty, and persuade Donald Trump to adopt Swedish parental leave policies. But each of these things is perfectly logically, nomologically, and metaphysically possible" (*Southwood 2018, 3*). The ultra-thin sense of binary feasibility does not capture the common way we use binary feasibility.[15]

Conceptually, such worries might be addressed by a more sophisticated elaboration of the account. However, as I aim for a conception that can guide our feasibility assessments while making sense of regular feasibility statements, the CPA seems ill-suited. The probabilistic understanding is misguiding as it ignores much of the complexities of the political world. This narrowing of feasibility goes back to three conceptual issues: a focus on clear agents, a focus on single

---

[14] This most naturally arises in the economic and technical sphere but is all too common in politics, where you might only have the political resources for one "expensive" reform. In some readings, for example, Barack Obama lacked the political resources to tackle health care reform and climate change in 2009 and opted for the former at the expense of supporting a more ambitious treaty at the COP15 in Copenhagen. Thinking of feasibility in the logical form above would not consider such considerations as part of feasibility.

[15] If we understand binary feasibility in terms of logical and akin impossibilities, it further loses all reference to *accessibility*. Logical and nomological consistency can be assessed without considering a trajectory at all. Consequently, Gilabert and Lawford-Smith (*2012*) applies this notion in their account of non-ideal theorizing only to principles, not states of the world. However, this binary sense of feasibility loses all traction on the "practicability" or "realizability" of a particular goal, something I take to be central to feasibility.

actions instead of a broader scope, and a marginalization of binary feasibility to an ultra-thin sense. We need an alternative understanding of feasibility that does justice to the complexities of mitigating climate change.

2.4 Feasibility as Restricted Possibility

We need an adequate conception of feasibility in relation to how we use the term and, which gives us appropriate guidance for assessing feasibility. I rejected the CPA account, as its narrow focus seems unhelpful and misleading for grounding feasibility assessments. There is an alternative in the literature in the Restricted Possibility Account by Wiens (*2015*).[16] It better captures the use of feasibility and provides a helpful explanation for scientific feasibility assessments. However, the RPA leaves us with two open questions: how to distinguish feasibility from mere possibility (the *threshold problem*) and how feasibility claims can guide us in deliberating what to do (the *guidance problem*).

While the CPA account marginalizes binary feasibility, the account proposed by David Wiens (*2015*) takes binary feasibility as the central meaning of feasibility. According to the *Restricted Possibility Account (RPA)*, an outcome is feasible if it is attainable given the resources and processes that are available to us. The RPA is an extension of the economic concept of the production frontier, which describes the attainable bundle of goods (for example, for a company) relative to the resources and production processes available. In order to capture feasibility, Wiens extends the framework of the production frontier to an "all-purpose production possibility frontier," which includes all different kinds of means for altering the status quo, including logical, physical, institutional, technological, economic, social, and motivational resources. Resource limitations along these different dimensions give rise to a "multidimensional constraint set" (*Wiens 2015, 453*), which delimits the space of a feasible state of affairs. According to the RPA, an outcome is feasible only if it is attainable given the resources (of all kinds) and conversion processes at our disposal.

Four elements are at the heart of his proposal: (1) *resources* are all means available for altering the status quo. Wiens has a fairly broad notion of resources, attentive to any fact "that constrains our capacity to alter the status quo" (*Wiens 2015, 452*). Such means include natural resources, machines, institutional capacities, economic resources, political means of influencing decisions, etc. Resources

---

[16] Southwood (*2019*) distinguishes a cost-based account as a third alternative but since there is no fully developed conception, I will only discuss these two contenders.

can be used for different ends but are typically finite and cannot be used unlimited times. Due to their finite nature, these resources translate to (2) *constraints* on the input of realizing a desirable state of affairs. Constraints arise because we lack the means to alter some aspects of the status quo. Importantly, we do not need to think of the sets of constraints given only by the current stock of resources. There are available (3) *conversion processes* between resources from one category into another. For example, we can sometimes increase agents' motivational resources through economic or other incentives. Skilled agents can turn one resource into another using their time and effort. For some resources, it is also possible to invest a given stock of resources and thereby increase it over time. Finally, (4) *causal processes* describe what we can do given the resource stock we attained. This is the element of human agency. Given these elements,

> "realizing a target state of affairs is feasible only if there is an attainable resource stock that enables us to realize it. [...] Colloquially, feasible states of affairs have production input demands that we can satisfy (now or in the future) given the all-purpose resources available to us; infeasible states are 'too expensive' given the set of all-purpose resource stocks that are attainable" (*Wiens 2015, 455*).

To judge whether a proposal is feasible, we thus need to analyze what resource stock would be necessary to realize it (given the causal processes) and whether we can get to the needed resource stock, given the current resources and processes of conversion at our disposal.[17]

The RPA provides a flexible framework that can guide scientific assessments and fits well with the term's common usage in this context. *First*, the RPA provides a good conceptual basis for analyzing feasibility concerning complex and large-scale problems. Feasibility is no longer restricted to the narrow logical form of being an agentive-modal. Getting to the attainable resource stock will involve actions of various agents but must not be restricted to it. Wiens account can thus capture feasibility claims that start from a non-agential perspective, as in the examples BERLIN2030 or "LÜTZI BLEIBT". This allows non-agential dynamics to play a meaningful role in making some outcomes more feasible. Nevertheless, feasibility assessments can be used to *disclose* the central causal processes by agents necessary to reach a goal. Further, RPA allows for a derivative sense of

---

[17] Wiens spells this out in terms of the logic of possible worlds, thus translating the modal terms attainability and accessibility further. Feasible states are states that exist in at least one possible world "circumstantially accessible from the actual world" (*Wiens 2015, 458*). However, we do not need to be concerned with this level of detail.

"feasibility for," if one constrains the analysis to the causal powers of a particular agent. Something is feasible for agent a if the state of affairs obtains in an accessible world "as a result of a causal process that involves actions taken by a" (*Wiens 2015, 460*).

Second, according to the RPA, feasibility demands providing pathways from the current state of the world to the outcome in question, paying attention to all kinds of constraints simultaneously. Feasibility is about finding *"viable pathways"*. Something is feasible if we can find such a trajectory, and what makes it feasible are the processes and resources usages involved in it. Binary feasibility is truly about accessibility, something many authors have pointed out as the term's central meaning (cf. *Cohen 2009*; *Erman and Möller 2020*).

Third, the RPA can accommodate considerations of *co-feasibility* explicitly. Given that we have dropped the agentive modal form, there is no longer a problem to include complex states of the world as an outcome. Analyzing such states, like the Paris Goals or achieving climate neutrality of some country or city, requires us to focus on aspects of co-feasibility and highlight trade-offs, interdependencies, and resource conflicts within these complex target states.

Fourth, the RPA fits well with how we think and use the term ordinarily. It captures the pathway character of feasibility and provides us with an analysis of binary feasibility. This is how most people use the concept. Most English dictionaries do not list comparative or superlative forms of feasibility at all (cf. *Collins English Dictionary 2023*; *Merriam Webster 2023*; *Vocabulary.com 2023*). A search in Google ngram, which lists occurrences in the vast digitalized library of Google, gives us very few appearances of "more feasible," "less feasible," or "most feasible." In contrast, "feasible" appears regularly throughout books in the 20th century (*Google Books 2023*). The New York Times archive lists 750 articles in the last decades that use "more feasible," compared to the 47.000 articles with the word "feasible" in its binary form (*New York Times 2023*). A Google Scholar search reveals a ratio of 1:20 in favor of the binary usage of "feasible." Also, the four anecdotes above suggest that feasibility's core meaning is binary.

The RPA drops the scalar sense of feasibility, and one might worry that we lose the seemingly helpful concept of a soft constraint. However, the distinction between soft and hard constraints is less clear on a closer look. According to Gilabert and Lawford-Smith (*2012*), soft constraints are constraints we could change over time. Such seeming constraints are part of an analysis based on the RPA in the form of processes of conversion and investment. For example, if a budget constraint is "soft" because we can change it by collecting more money, the RPA would reveal the necessary trajectory to involve this kind of budget extension. It would not appear as a proper constraint. However, this seems to get things right. Soft constraints seem perplexing on a closer look. As they are changeable, why call them constraints at all? After all, most trajectories towards a desirable goal will involve changing some aspects of our environment. Climbing a mountain might involve the constraint that I will injure my feet, but this is changeable by putting on proper boots. Are my feet a soft constraint? Is the empty tank of my car a soft constraint, although I can change it easily by fueling up? Soft constraints might be understood as picking out some aspects of trajectories, flagging them as central steps for reaching a goal. However, such *highlighting* could also be done explicitly without labeling them a constraint.[18]

As the RPA provides a good explication of feasibility that we can rely on for assessing feasibility, let me describe the implications of the RPA for feasibility assessments. On the most general level, the RPA teaches us to find *viable pathways*. Viable pathways are trajectories from the status quo to the outcome that do not violate any constraint. These constraints can be analyzed, as it is the common practice in assessments, along *different "dimensions" of feasibility*. We can differentiate between economic, institutional, technological, social, geophysical, cognitive, biological, political, and other kinds of feasibility. These dimensions cluster different feasibility facts within the boundaries of well-established scientific disciplines. These boundaries will not always be sharp, but I consider it specific enough to use these dimensions. They establish what we may call *specific feasibility*.

Such clustering can help specify the meaning of our feasibility claims. Sometimes, we are interested simply in one kind of feasibility, either because we already have a good grasp on the other dimensions of feasibility (for instance, a firm might need to know whether building an established piece of technology is economically feasible) or since such a dimensional feasibility claim is all we can hope for at the moment (for instance investigating whether using algae for capturing CO_2 is ecologically feasible). Moreover, specifying more clearly what kind of limitation gives rise to a particular claim of infeasibility is more fine-grained information than claiming that an outcome is infeasible.

Dimensions of feasibility broadly group together "related facts about the means at our disposal for altering the status quo" (*Wiens 2015, 453*). What kind of grouping is appropriate is an epistemic question and depends on what kind of problem we are facing. The IPCC provides six relevant dimensions of feasibility for the climate mitigation context: "geophysical, environmental-ecological, technological, economic, socio-cultural and institutional factors that enable or

---

[18] We can also fix certain constraints with the RPA, as one can ask whether an outcome is feasible within the given budget, for example. That is an explicit scenario.

constrain the implementation of an option" (*IPCC 2022, 1802*; *Masson-Delmotte et al. 2018, 381*):

- *Geophysical feasibility* refers to constraints and possibilities arising from factors related to the physical aspects of the climate and earth system, the physical potential of specific resources, or the available land. Relevant examples of geophysical feasibility are the available carbon budget for a given temperature goal or the physical potential for wind, solar, and biomass production.
- *Environmental-ecological feasibility* generally concerns aspects of interactions with ecosystems and environmental side-effects. The AR6 gives air pollution, toxic waste, water availability, and biodiversity issues as indicators for feasibility in this dimension (*IPCC 2022, 1837*).
- *Technological feasibility* addresses the availability of the technologies necessary to perform specific measures, e.g., technological development, its potential to be upscaled, and how easy the technologies are to be implemented.
- *Economic feasibility* is often understood in terms of overall costs, though different kinds, timings, and distribution of costs also affect the economic feasibility. Further indicators of economic feasibility are macro-economic effects, e.g., "employment effects and economic growth" (*IPCC 2022, 1837*).
- *Socio-cultural feasibility* refers to factors arising from the social world, such as "public acceptance" and effects on socially relevant categories, such as health, well-being, and "equity and justice across groups, regions, and generations, including security of energy, water, food and poverty" (*IPCC 2022, 1837*).
- Finally, *institutional feasibility* addresses issues related to the availability of agents, their capabilities, and resources, as well as whether there is institutional support for specific options (*IPCC 2022, 1837*).

Feasibility can be assessed along these dimensions. Looking at them in isolation provides us with specific feasibility judgments. However, what is normatively consequential in the abovementioned sense are *overall* or *all-things-considered feasibility judgments*. When deliberating what to do, only *overall infeasibility* can rule out proposals in our practical deliberation.[19] Such judgments can only be

---

[19] Confusingly, the term "political feasibility" has been used in both the specific sense (*Majone 1975*) and the general sense (*Gilabert and Lawford-Smith 2012*). In the specific sense, "political feasibility" refers to constraints arising due to political considerations, for example, lack of public support or institutional barriers (cf. *Patterson et al. 2018*). This is how I will use the term as well. In the general sense, "political feasibility" refers to feasibility considerations relevant in the context of normative *political* theory, thus, in evaluating what

attained when we have considered all relevant dimensions in conjunction, and the pathway towards the outcome does not violate any constraints.

The RPA implies the need for an integrated analysis to gain such judgments of overall feasibility. As we often can transform resources from one dimension to another, it is insufficient to look at the dimensions in isolation and add them up. If we lack the economic resources to bring some outcome about, this economic infeasibility does not automatically translate into overall infeasibility. We might be able to attain the resources by using other resources. Moreover, even if an outcome seems feasible along all dimensions looked at in isolation, it might still be infeasible overall. For instance, we might use the same economic resources to incentivize people to compensate for lacking motivational resources *and* to gain the necessary technological developments through investments. As we cannot use one resource for both, the RPA will lead us to conclude that the outcome is infeasible. The RPA implies the necessity of an overarching perspective in answering feasibility claims. A viable pathway is one that does not violate any constraint when looked at in conjunction.[20]

Finally, I need to say more about how the RPA relates to agents' willingness. This is a difficult topic that has concerned the conceptual debate (cf. *Roser 2016*; *Stemplowska 2020*). Wiens himself includes motivational constraints into his analysis of feasibility when they "identify the limits of what people can be motivated to do given intrinsic features of human agents that affect motivation (including affective biases, prejudices and fears), as well as the extrinsic features of an agent's environment that interface with her intrinsic motivational capacities (including social norms and incentives)" (*Wiens 2015, 453*). In other words, when we can use other resources to alter an agent's motivation, including our own, the options made available by such measure are part of our feasible set. What we cannot incentivize ourselves and others to do is beyond our motivation and thus infeasible for them or us. This seems to get things roughly right. Importantly, what is said does not preclude us from distinguishing morally dubious with proper motivational constraints. If I am a lazy cook, cooking dinner for my roommates is within the limits of what I can be motivated to do (given the intrinsic and extrinsic features), even if I routinely fail to do so since I cannot

---

is the best thing to do in terms of justice and other political values. Here, then, "political" is denominating the overarching perspective.

[20] The small-scale examples with clear agents and actions used by many authors in the feasibility debate pretend that we already have such a perspective. In reality, though, this overarching perspective is what is so difficult to achieve for determining feasibility claims. It will involve knowledge from different perspectives, as the nature of the involved "resources" and "processes" is very wide and thus is responsive to very different kinds of expertise.

be bothered. Where to draw the line between those features that count towards feasibility and those that are illegitimate due to them going back to unwillingness is an open question. *That* we need to draw this line, though, is not in doubt in the RPA. I will, thus, rely on this distinction even though I cannot give a more systematic analysis of this aspect at this point.

However, while more attuned to large-scale social outcomes, there are short-comings with the RPA on a conceptual level. In dropping the scalar notion, Wiens's account becomes vulnerable to being too permissive. As Wiens read-ily admits, the RPA only gives us a necessary condition of feasibility, being, in an important sense, incomplete. Southwood (*2018*) points out that the RPA is vulnerable to counterexamples in the form of "counterfactual flukes," states that only come about by highly improbable instances of luck. Southwood explains how "there is obviously at least one world at which' I win 10 different lotter-ies in a single day 'that is circumstantially accessible from the actual world' (*Southwood 2018, 4–5*). Alternatively, consider the example above. There is a pathway from the status quo to me buying a plane involving playing the lottery and winning. This is practical as it involves actions by me, relies on resources and conversion processes (buying a lottery ticket), and there is a possible path for reaching the outcome. The RPA needs to call such outcomes feasible, but they clearly are not. We are confronted with the problem of *defining a threshold*, at which mere practical possibility turns into feasibility.[21] How much reliance on chance is permissible for something to be feasible? The problem of a threshold when some outcome is feasibility instead of being merely possible thus remains an open question.

Moreover, I outlined that feasibility could play an important role in *guiding action*. The CPA gave us an answer on how this can be understood, namely in ranking alternatives with respect to their probability. Wien's account lacks such an answer. Thus, it must be accompanied by an explanation of how feasibility assessment can be relevant beyond the binary role of ruling out specific proposals. We can answer these limitations if we accept the value-laden nature of feasibility. Accepting value-laden feasibility will help to answer the two open questions of *defining a threshold* and *guiding action*, or so I would like to argue in Chap. 3. The RPA provides a good conceptual framework for scientific assessments of feasibility. I will summarize it below.

---

[21] Defining a threshold would also arise in the CPA as it is described by Brennan and South-wood (*2007*), who proposed that a state of affairs is feasible for an agent (or a set of agents) if it would be "reasonably likely" that they succeed in realizing x if they tried. What counts as "reasonable" depends on such a threshold.

2.5 Summary

Before moving on, let me briefly summarize my discussion. I described that the most well-known approach, the *Conditional Probability Approach*, which takes feasibility as a function of the probability of success, is tailored too closely to neat examples with clear agents and action. Probabilistic feasibility is unhelpful and potentially misguiding in assessing feasibility in the complex terrain of climate change. Further, the CPA does not capture the common binary usage of feasibility well.

I argued that the *Restricted Possibility Account* brought forward by Wiens does better in several aspects. It gives a precise analysis of binary feasibility in terms of accessible pathways that respect the complexities of large-scale problems. The non-agential starting point of feasibility can capture aspects of co-feasibility and dynamics, which lack clear agents. However, it leaves two central elements of feasibility that need to be answered. Feasibility must be more substantive than mere possibility. However, Wien's account lacks a way to differentiate between feasible and merely possible pathways, for instance, due to trajectories that involve unreasonable instances of luck. I called this the *threshold problem*. Any account of feasibility must give some answer to how to distinguish whether the outcome is feasible instead of merely possible.

Further, feasibility is often used as guiding action, for instance, in comparative evaluations of pathways and goals as proposed by the CPA. We sometimes speak of an outcome being less realistic or more challenging, and this evaluation matters to us. The answer of the CPA to this question was straightforward: feasibility assessments can "rank" proposals in light of the conditional probability of success. The RPA did not provide an answer.

The RPA, though, gives a good explication of the general characteristics of feasibility. Most generally, feasibility is about finding viable trajectories from the status quo toward a desirable goal in the future. In the complex terrain of climate mitigation, this will involve assessing multiple pathways and paying attention to interdependencies and trade-offs that could affect the feasibility of some outcome. Only if all constraints are considered an outcome is feasible.

The Value-Dimension of Feasibility

3

This chapter will argue that we should accept a value dimension to feasibility. I will argue to understand feasibility as thick in three ways: feasibility judgments involve value judgments concerning unacceptable means, side-effects, and uncertainties. If we accept this, the two open challenges the Restricted Possibility Account, as proposed by Wiens, left us with can be answered. Thick feasibility can explain how feasibility is more substantive than the merely practically possible, and it explains how feasibility judgments are, at times, guiding our practical thinking beyond ruling out proposals. This chapter further specifies that the normative role of feasibility is to rule out proposals in practical deliberation, implying that we should be cautious not to rule out pathways or goals too prematurely.

The first section introduces the widespread presumption of *"Descriptive Feasibility"* and why it is so attractive. I will then survey arguments for a value dimension to feasibility from the literature (Sect. 3.2). I will present my own arguments for accepting such a dimension (Sect. 3.3) and explain what this value dimension consists of (Sect. 3.4). Finally, I will return to the normative role of feasibility in light of the thickness of the concept (Sect. 3.5).

3.1 The Promise of Descriptivity

There is a widely held assumption in the conceptual and empirical literature, which I will call *"Descriptive Feasibility"*. It holds that feasibility is an empirical term that is conceptually free of normative content. Such a conception implies that

© The Author(s) 2025
S. Hollnaicher, *Assessing Feasibility with Value-laden Models*,
https://doi.org/10.1007/978-3-662-70714-2_3

feasibility assessments, properly understood, cannot depend on value judgments. This section will introduce this presupposition.

For instance, the two main accounts of feasibility in the philosophical literature discussed in Chap. 2, share this presupposition. Wiens writes: "It is worth pausing to note that this analysis of feasibility is void of any moral content. Accordingly, feasibility assessments do not incorporate our judgments about which states of affairs are worth realizing from a moral standpoint" (*Wiens 2015, 9–10*). He goes on:

> "A non-moralized analysis of feasibility gives us a clear view of the role played by particular facts in our feasibility assessments and permits us to see clearly at which points in our normative analysis our feasibility judgments, as opposed to our desirability judgments, are doing the work. A moralized notion of feasibility precludes this analytic separation" (*Wiens 2015, 9*).

Many philosophers working on the meaning of feasibility share this view (cf. *Brennan and Southwood 2007*; *Cohen 2009*; *Gilabert and Lawford-Smith 2012*). Gilabert and Lawford-Smith claim "that it is possible to distinguish considerations about what is feasible from considerations about what is desirable. When we want to know how feasible X's bringing about O in Z is, it is possible to assess dispassionately the hard and soft constraints upon X's φ-ing in Z" (*Gilabert and Lawford-Smith 2012, 816*). Value-*laden* feasibility, this quote implies, would be a "passionate" analysis, which risks letting one's ethics influence feasibility facts. They go on: "We deny that any moral paradigm has the metaphysical standing to make it the case that a given proposal cannot be realized" (*Gilabert and Lawford-Smith 2012, 817*).[1]

One can understand *Descriptive Feasibility* as a proxy for a more fundamental *dichotomy* between the desirable and the feasible (the normative and the descriptive, or between values and facts). Separating these two kingdoms of judgments has a long tradition in philosophy, for instance, established by Hume's law that one cannot derive an "ought" from an "is" (*Hume 1739--40 [2001]*) or by Kant's analysis that as long as moral aims are not "demonstrably impossible to fulfill [...] [they] amount to duties" (*Kant and Wood 1996*). Ethical judgments cannot

---

[1] The CPA of Gilabert and Lawford-Smith (*2012*), however, took the scalar sense of some outcomes being "more feasible" than an alternative to be more central. A slight rhetorical move here is to only speak of binary feasibility in the final sentence. The authors take binary feasibility to be only concerned with logical and physical impossibilities and thus more easily understandable as value-free. They, though, must mean to extend the value-free analysis to their scalar notion of feasibility as well.

be deduced from pure facts. Vice versa, feasibility facts cannot be dependent on what we think should be the case, ethically speaking.

This dichotomy is also the ground on which we sometimes envisage a division of labor. This division holds that moral and political philosophy, religion, politics, and other sources of values are authorized to discuss the moral quality of an ideal, all the while the sciences provide the feasibility facts of how we get to the ideal or how attainable it is. A. John Simmons, for example, interpreting Rawls' conception of a nonideal theory, describes such a distinction of responsibility:

> "Although much is obviously left vague here, we can at least infer this much from Rawls's remarks: nonideal theory will require judgments of both philosophical and social-scientific sorts. Determinations of a policy's 'moral permissibility' obviously lie in the proper domain of moral and political philosophy, as do judgments of grievousness, which depend on prior ideal theorizing. Determinations of 'political possibility' and 'likely effectiveness,' on the other hand, seem more naturally to require the expertise of, e.g., political scientists, economists, and psychologists" (*Simmons 2010, 19*).

Similarly, Adam Swift describes that the job of ethicists is to rank "the options that social science tells us to be within the feasible set" (*Swift 2008, 369*). This neat division of labor relies on the promise of Descriptive Feasibility and, subsequently, a belief in a value-free science, which can provide feasibility facts relevant to political theory and proposals but independent of normative judgments (cf. *Lenzi and Kowarsch 2021*).

Modelers and social scientists often buy into this presumption of Descriptive Feasibility and use it to demarcate their work from value-laden enterprises as normative theories or actual policymaking. For instance, Brutschin et al. "stress the importance of a conceptual and operational distinction between feasibility and desirability" to highlight their method of determining feasibility claims being putatively empirical and evidence-based (*Brutschin et al. 2021, 2*). The framing in IAM papers of assessing only the "feasibility" of climate goals and pathways is often used to demarcate their work from value aspects. Gambhir et al. (*2017*) explore "the critical notion of how feasible it is to achieve long-term mitigation goals to limit global temperature change" and stresses that they exclude "political and social concerns" (*Gambhir et al. 2017, 2*). P. C. Stern et al. (*2023*) call for "[s]cientifically grounded feasibility assessments," presumably to separate them more clearly from value aspects.

Framing one's work under the term "feasibility" is a proxy for a more general vision of legitimate science: "When defining feasibility, actors are speaking to what they see as the proper relationship between modeling and climate policy in particular, and science and society writ large" (*Low and Schäfer 2020*, 6). Low and Schäfer describe the "incumbent perspective" of modelers, which stresses the independence of inquiry from politics and takes value questions to arise mainly in communicating results. In this perspective, a clear separation of science and value-laden policymaking is possible. Modelers see it as their explicit task and mandate to only assess the feasibility of mitigation options and pathways, refraining from judging their desirability.

References to feasibility for delineating objective scientific advice can already be found in the work of Max Weber and his discussion of the value-free ideal. In arguing for the possibility of objective social science, Weber (*1904*) draws on the distinction between means and ends. Weber argues that social scientists' work should only answer three questions: whether specific means are adequate for a given goal, what the implications of different means are, and lastly, what the ends mean and whether they are consistent (*Weber 1904, 25*). According to Weber, scientists have no business guiding whether to adopt certain ends. However, the three tasks imply that scientists can assess the feasibility of these ends and guide us in achieving them. In doing so, their work can be free of value judgments in the relevant sense.

The conceptual presumption of *Descriptive Feasibility* appears attractive to scientists and philosophers, as it promises to ground a clear boundary between the desirable and the feasible. Many contributions define feasibility as independent of value judgments. Descriptive feasibility is a central commitment of the two main accounts. However, not all conceptual discussions on feasibility buy into Descriptive Feasibility. The following section will review some contributions that see feasibility as inherently value-laden.

3.2 Existing Arguments for a Value Dimension

The previous sections described the shared assumption among conceptual analysts, social scientists, and modelers alike of feasibility being void of normative content. I called this *Descriptive Feasibility*, which states that judgments of feasibility do not involve or depend on value judgments. Descriptive Feasibility is widespread in philosophical accounts of feasibility (cf. *Majone 1975*; *Brennan and Southwood 2007*; *Gilabert and Lawford-Smith 2012*; *Lawford-Smith 2012*; *Wiens 2015*; *Southwood 2018*; *Stemplowska 2020*). However, there have also been some contributions that argue for a value-laden concept of feasibility. None of these contributions gives a fully worked-out conception of feasibility, but they lay the grounds to explore how feasibility conceptually depends on value judgments.[2]

Juha Räikkä (*1998*) introduces an explicit argument for accepting a value dimension to feasibility. Räikkä argues that the feasibility of an outcome also involves the moral costs of transitioning towards the ideal. Calling some ideal feasible, in his view, means that it is achievable without undue or unjustifiably high costs. The argument is the following: sometimes, we must dismiss an ideal, not because of its inherent moral value, but because the costs of achieving it would be too high to accept. Räikkä does not provide an example of his own, but we might think of the following: It seems to be a desirable ideal that Germany is fossil-free in 2030. However, imagine that the only way to achieve it would be by eco-authoritarian means, having strong police enforcing strict resource uses. If so, the ideal of climate neutrality by 2030 must be dismissed (as an ideal in political theory).

If so, there are three options to explain how this can be the case. Either the ideal has become less desirable due to the problematic means of achieving it. However, that seems implausible. Alternatively, the "feasibility approach" and the "desirability approach," as Räikkä calls them, are not the only ways to make political theories. Perhaps there is a third dimension, though it is unclear what this would look like. Räikkä finds a third option most plausible. The *feasibility* of the ideal should involve the moral costs of transitioning towards the ideal, and, therefore, we should think of this ideal as not feasible (and thus reject it).

This third option would imply accepting a value dimension to feasibility. This option finds further support as one might "say that a social ideal is not feasible, even if it is capable of being implemented, that is, even if it is possible" (*Räikkä 1998, 36*). This difference between mere possibility and feasibility might be grounded in a normative dimension of feasibility. An outcome is feasible if it is "possible and acceptable to carry [it] out when the necessary costs of changeover are taken into account. [...] [I]t becomes partly a normative matter to decide which institutional arrangements are feasible and which are not" (*Räikkä 1998, 37*). This provides us with a realistic rendering of the feasibility issue, as it alerts to the frequent encounters of attributions of infeasibility being "based on normative views (regarding the necessary moral costs of changeover), not simply on views concerning literal possibility and impossibility" (*Räikkä 1998, 39*).

---

[2] These discussions foremost concern the role of feasibility within political theory. Some, for instance, distinguish between feasibility constraints in political theory and political feasibility as it applies to politicians or other practitioners (*D. Miller 2013, 36–38*; *Räikkä 1998, 28*). I take scientific advice, as done in IPCC reports and similar reports, to be sufficiently close to the role of feasibility constraints in political theories.

A similar argument can be found in Allen Buchanan (*2003*), who argues that beyond feasibility (in an ultra-thin sense), further accessibility and what he calls "moral accessibility" are critical aspects of a political theory: "Other things being equal, a theory should not only specify an ideal state of affairs that can be reached from where we are (though perhaps only after a laborious and extended process of change), but also the transition from where we are to the ideal state of affairs should be achievable without unacceptable moral costs" (*Buchanan 2003, 61*). A theory involves unacceptable moral costs of transition if any path towards its realization from where we are will bring up "other, comparable evils" (*Buchanan 2003, 62*). Thus, nonideal theorizing needs to assess when the costs are unbearable high, which becomes part of feasibility. This notion of "unbearable" costs will depend on the stakes involved.

David D. Miller (*2013*) argues that value aspects are a common feature of actual political theory. Rawls, for instance, took certain kinds of social institutions so deeply ingrained in our society that political proposals that conflict with them are to be considered infeasible (cf. *E. Kelly and Rawls 2001, 165*). Miller writes in his political theory "for earthlings":

> "I believe in fact that the notion of practical possibility that Rawls relies upon has an inescapable normative element. The limits of political possibility are set not just by physical and sociological laws, but by implicit assumptions about what, for us, would count as a tolerable or intolerable outcome" (*D. Miller 2013, 32*).

Rawls takes the existence of a family (though not a particular form of a family) to be so fundamental to our society that principles of justice, which are inconsistent with this institution, must be seen as infeasible (cf. *E. Kelly and Rawls 2001, 165*).

Miller argues that such a limit of possibility is not factual but normative, but it is nevertheless an essential part of how we construct political theories. Other such value commitments as normative side-constraints of feasibility could be assumed, for instance, in inviolable democratic norms or a certain level of equality or social justice. If this is true, he argues, "we need to be clear about exactly what we are taking for granted when we assert the principles in question" (*D. Miller 2013, 35*). That is, we must make normative commitments we consider part of the feasible limits, explicit as best we can.

Finally, Alan Hamlin (*2017*) discusses value feasibility as one underappreciated aspect of feasibility which, if accepted, would provide a "more realistic rendering of feasibility" (*Hamlin 2017, 211*). Value feasibility is the "feasibility of realizing particular values or combinations of values, or of achieving outcomes while respecting particular values" (*Hamlin 2017, 215*), for instance, asking if "a particular outcome is feasible while respecting particular norms (for example, treating animals appropriately)" (*Hamlin 2017, 215*). Such questions (implicit or explicit) are prevalent questions of feasibility but cannot be answered solely relying on empirical constraints. Hamlin's characterization contains two conceptions of value feasibility that we should keep distinct. The first, "feasibility of realizing particular values," is simply the feasibility of a specific desirable outcome. Such a notion is not distinctively value-laden, as it only concerns a more complex social outcome than a concrete proposal. The second sense, "achieving outcomes while respecting particular values," is more akin to the normative constraints on feasibility we find in the contributions by Buchanan, Miller, and Räikkä. For example, Hamlin writes: "[T]reating animals appropriately may constrain our ability to produce some outcome that would otherwise be feasible if animals were treated as mere resource[s]" (*Hamlin 2017, 216*). Normative considerations serve here as "side constraints" and thus could be conceptualized along with other constraints, such as economic constraints, limited institutional capacities, or limited human skills. Such (implicit) side constraints turn feasibility into a partly normative concept.

These arguments suggest a value dimension to a realistic view of feasibility, which the philosophical analysis of the concept tends to miss. What is feasible in this sense is attentive to normative aspects of the path of transition, either by excluding truly immoral means (*Buchanan 2003*), respecting deep value commitments of society (*D. Miller 2013*), or being sensitive to acceptable costs and means (*Räikkä 1998*; *Hamlin 2017*). The arguments differ in the form of these normative constraints, whether value aspects are objective or subjective, and whether value feasibility is gradual or binary. The common core is that feasibility depends in essential ways on value judgments, thus rejecting Descriptive Feasibility. In the following section, I will give further support in favor of a value-laden conception of feasibility. Sect. 3.4 will then explain this normative dimension fully.

3.3 In Favor of Thick Feasibility

If we look at typical examples of feasibility judgments, they very often depend on implicit value-laden assumptions. This section lays out the arguments for feasibility being a *thick concept*, that is, a concept that involves descriptive and normative aspects in a closely entangled way.

A thick conception of feasibility takes value judgment to play a substantive role in determining the meaning and application of feasibility claims. Generally, the term "thick concepts" applies to concepts that combine descriptive and normative aspects in an entangled fashion. The thickness of concepts is often explained by reference to virtue concepts such as "braveness." Being brave, on the one hand, describes certain traits and behaviors of the person in question. However, it does not merely describe but also evaluates these traits and behaviors as positive. There are, of course, a range of open questions concerning how this is to be understood more concretely, for instance, whether this is a pragmatic or semantic thesis or concerning how deep the entanglement goes (cf. *Väyrynen 2013*; *Kirchin 2013*). Recently, the role of thick concepts in science has received greater attention as well (cf. *Möller 2012*; *Shockley 2012*; *Alexandrova 2017*; *Reiss 2017*; *Djordjevic and Herfeld 2021*). Dupré (*2007*) argues that thick concepts are essential in tying science to our interests and values. Thick concepts thus play an important role in ensuring that scientific knowledge is relevant, being about things we care about.

Bracketing these questions for the moment, I will rely on the definition of Elizabeth Anderson: "A concept is thickly evaluative if (a) its application is guided by empirical facts; (b) it licences normative inferences; and (c) interests and values guide the extension of the concept (that is, what unifies items falling under the concept is the relation they bear to some common or analogous interest or values)" (*E. Anderson 2002, 504–5*).[3] For discussing a thick conception of feasibility, the most crucial task is to explicate the value dimension of feasibility, that is, in Anderson's terms, the way feasibility "(b) licences normative inferences" and the way "(c) interests and values guide the extension" of feasibility. Aspect (c) is the task of the following section—(b) is addressed in Chap. 5. Before doing so, let me give the arguments in favor of feasibility's thickness.

Take, for instance, the statement cited by Wiens in his conceptual analysis of feasibility: "Former Indian civil servant Prodipto Ghosh's claim that 'it is not feasible [for India] to do anything [about climate change] at this stage [beyond voluntary reductions of greenhouse gases]. With the present state of technology development, we are likely to encounter severe constraints to our growth'" (*Wiens 2015, 450*). Wiens considers this statement a proper feasibility judgment (note that this is compatible with Gosh being mistaken). A natural reading of such a statement is that pathways in which India decarbonizes faster are not feasible

---

[3] Bernhard Williams introduced the term thick concepts. He describes such concepts as both *world-guided* and *action-guiding*. The correct application of the word depends on facts of the world, but at the same time, the concept involves reasons for actions or evaluation of a situation (*Williams 1985, 140–41*). Alexandrova (*2017*) picks up on these discussions in proposing to understand claims on well-being as *"mixed claims"*. These empirical claims about causal or statistical relations contain one variable whose definition depends at least partly on a non-cognitive value judgment (*Alexandrova 2017, 82*).

because they imply socially and morally unacceptable burdens for the Indian population. For instance, Ghosh might think such pathways would not do enough to alleviate the extreme poverty in his country. So understood, this statement involves empirical judgments (e.g., the causal claims on how decarbonization can be achieved) *and* normative judgments (on what kinds of burdens would be unacceptable). Ghosh's statement makes the value-laden assumption that a certain reduction in growth would be "too severe" to accept.

When looking more closely at other examples of feasibility claims, such assumptions on unacceptable means seem hard to avoid in general. Whether one judges a stricter climate goal for Berlin to be feasible, as in the case of BERLIN2030, will also depend on implicit normative assumptions concerning which means one deems acceptable. For instance, most people's (including expert's) judgment on whether it is feasible for Berlin or Germany to reach climate neutrality will implicitly assume that the means applicable to such a judgment must respect fundamental values such as a democratic order, social stability, or avoiding too severe social burdens for people experiencing poverty. Surely, if pressed, people might admit that other ways of realizing climate neutrality are perfectly possible. While this could be understood as correcting their statement, a more natural reading is to see it as a slight change of subject. If we accept that feasibility is more substantive than mere possibility, then this slight difference in wording (from feasibility to possibility) retracts from common usages of feasibility. So, a first strand of support for thick feasibility is that many actual feasibility claims, as made in the policy discourse, involve some judgment on acceptable means and costs of achieving the outcome (cf. *Räikkä 1998*).

This challenge of distinguishing the feasible from the merely practically possible was one of the conceptual challenges neither of the two value-free accounts could answer. Value judgments can fill this gap. Feasibility is more substantive than possibility because feasibility judgments involve additional constraints and thresholds, which can only be applied with the support of value judgments. As I will explain below, there are specific ways in which values can play a role in feasibility judgments, and these roles for values help explain the substantiveness of feasibility in contrast to pure possibility claims. Moreover, accepting value dimensions in feasibility claims helps explain feasibility claims, which evaluate different pathways or goals in relative feasibility, for instance, the "degree of challenge" or the "level of realism." Such related terms might be accepted more readily as thick concepts, often used to describe essential aspects of feasibility assessments. This will explained in Sect. 3.5.

The following section will develop a thick conception of feasibility. It is an extension of the RPA account, though the normative dimension might also be compatible with other accounts.

3.4   A Thick Conception of Feasibility

This section explicates a *thick conception* of feasibility by extending the Restricted Possibility Account from Chap. 2. It proposes three ways value judgments are involved in feasibility claims: value judgments *(1)* exclude means that are deemed unacceptable in themselves or involve *(2)* unacceptable side-effects, and value judgments *(3)* determine a threshold concerning the acceptable level of uncertainty in feasibility claims.

To recall, the Restricted Possibility Account understands feasibility as the accessibility of a particular outcome from the status quo. In order to realize something, we need to use different resources, for instance, economic resources, technical skills, institutional capacities, or all other kinds of "all-purpose resources." Resources are typically finite, giving rise to different kinds of constraints. Whether an outcome is realizable depends on causal processes by agents who can use resources to bring about some desirable outcome. Importantly, though, Wiens *(2015)* describes that feasibility is not only about what we can do based on the current stock of resources but also what we can do given all "attainable" resource stocks. Through investing, conversing, or saving resources from where we are, different resource stocks are attainable in the future, making it possible to realize different outcomes. In short, an outcome is feasible if "there is an attainable resource stock that enables us to realize it" (*Wiens 2015, 455*). In less technical terms, something is feasible if there is a trajectory from the status quo toward the outcome in question, which simultaneously respects all constraints. Something is feasible if there is a viable path.

What counts as a "viable" path involves value-laden assumptions, or so I would like to argue, extending Wiens' analysis and challenging the value-free commitment of his account. Let us start with a simplified example: We as a group need to decide how to cross a steep and dangerous valley. Jabari, who knows this region better than the rest, says that a direct crossing is infeasible for us. However, he claims that there is a different path to get to the other side. It crosses the river further upstream and, though steeper, avoids the sharp rocks of this part of the river. Thus, Jabari claims that it is feasible for us to get to the other side (there is at least one viable trajectory).

To determine this, he made value judgments in at least three ways: *first*, he must have weighed the risk factors, determining which risk was acceptable to the group. In deciding what counts as too dangerous, he also must have weighed different dangers, for instance, how more significant chances of minor injuries compare to smaller chances of fatal accidents. *Second*, he must have made some preselection of acceptable means. For instance, it might have been possible to call the mountain rescue service and ask them to bring us to the other side, but the ethics of being a mountaineer does not allow to call them in circumstances that are not an emergency. Using all of the group's financial budget to call for a private helicopter service was also possible. However, Jabari did not consider it, as it appeared too unreasonable to take into account. Jabari (perhaps without considering them explicitly) excluded both these means beforehand. *Third*, Jabari weighs the uncertainty of actually getting to the other side, presumably taking into account how important it is for us to achieve that outcome. There is certainly some probability that the crossing fails. Whether it is certain enough to claim the crossing to be feasible (and not dismissing the option to try it at all) involved some value-laden decision on the uncertainties involved. So, Jabari made a range of value-laden decisions in reaching his feasibility judgment.

These aspects generalize in a thick conception of feasibility. As explained above, in a thick concept, "interests and values guide the extension of the concept" (*E. Anderson 2002, 504–5*). I propose the term feasibility depends in three ways on value judgments. The first is the following:

(1) **Acceptable Means:** Feasibility judgments involve value judgments concerning what means are acceptable.

Value judgments can play a role in preselecting the means that inform feasibility judgments. Often, some means are dismissed in the search for trajectories because they are deemed "unacceptable." Feasibility judgments often implicitly assume that one should not even consider certain means for realizing an outcome. Jabari, for instance, excluded the direct path as it would have resulted in a high risk of severe injuries. Moreover, he excluded the option to call the rescue team as unacceptable. Making a feasibility judgment *postulates* that the value judgment used to dismiss specific means is uncontroversial. However, feasibility claims of course sometimes wrongly exclude means as unacceptable. We can disagree on the normative judgments involved.

The most straightforward case of such a value judgment is in the exclusion of immoral means. We often judge something as infeasible even though we know that some odd pathway makes it possible, which, however, would violate moral rules. It might be possible to vaccinate your committed anti-vaxxer uncle by secretly anesthetizing him. However, it would be odd if someone considered this option and stated that him getting vaccinated was feasible. We would further think that such a person makes a moral mistake, suggesting that there is a value judgment. Unacceptable means can also be traced back to other value-laden aspects. Morality is a particularly clear case of value judgments in feasibility.

Means could be deemed socially unacceptable if they violate deep value commitments. For instance, suppose somebody claims that it is infeasible to avoid all unequal chances for children in terms of educational success, where the person implicitly dismisses some unacceptable measures, such as taking children away from their parents right after birth and raising them all in foster homes. This might be seen as socially unacceptable. Conventional or social value judgments are arguably also involved in other feasibility claims such as the expert claims made in BERLIN2030, where experts might have excluded means due to them conflicting with the deep value commitments of the citizen, for instance, excluding the option to ban cars entirely from the city. The common core is that some means are unacceptable for normative reasons. Applied to Wiens' account, feasibility judgments involve a value-laden preselection of what kind of resource, conversion processes, or causal processes are acceptable to justify judging something as feasible.

The second value aspect of feasibility claims is the following:

(2) Acceptable Side-Effects: Feasibility judgments involve value judgments concerning what side-effects are acceptable.

Sometimes, not the means themselves are unacceptable, but their side-effects. Means can be economically unacceptable if they lead to high costs or ecologically unacceptable if they have harmful side effects for nature and nonhuman animals. For instance, take Ghosh's statement above that it is infeasible for India to do more about climate change because they would otherwise encounter unacceptable economic effects. Nothing about the means of mitigating climate change is in themselves seems unacceptable. Decarbonizing the energy, transportation, and other systems is by itself acceptable. However, the implications of these measures, given the technological and economic status of the time, were deemed too severe. This is a judgment on acceptable side effects of specific pathways and measures.

Taking (1) and (2) together, we can say that an outcome is feasible if there is a resource stock that enables us to realize it, which is attainable without relying on unacceptable means and without producing unacceptable side effects.

Further value judgments arise because we have to evaluate how sure we need to be concerning internal and external uncertainties in judging an outcome to be feasible:

(3) Threshold uncertainty: Feasibility judgments involve value judgments concerning the acceptable level of uncertainty.

Most of the outcomes we can achieve will depend on some luck. Even skilled people cannot guarantee the success of their actions. We all need to depend on the world to be (at least minimally) supportive of our actions to have some chance of succeeding. Feasibility takes this warranted level of trust in the world into account. If the way to the restaurant is suddenly flooded or all employees are on sick leave, I will not succeed in getting my lunch. Nevertheless, it would be odd to think that it was wrong to consider it feasible beforehand due to this remote risk. Complex social outcomes such as achieving climate neutrality will involve many sources of uncertainty. For instance, we might count on the negative emissions from reforesting a specific area to achieve climate neutrality. Nevertheless, of course, this can go wrong in many ways. A fire could burn down the forest, or some new species of bug could kill all the trees. However, such uncertainties must not make the pathway infeasible. Some uncertainties in complex pathways might even be "deep," thus not allowing us to assign any probability to them (*Frisch 2013, 120*).[4] Judging some outcome to be feasible involves determining some level of sufficient certainty.

An outcome is feasible only if a resource stock is attainable *with sufficient certainty* (without violating core value commitments), enabling us to *reliably* realize it.[5] Whatever these terms mean more precisely, any evaluation of what counts as "sufficiently certain" and "reliably" will involve value judgments. The threshold for certainty will arguably be higher in cases where the stakes are more serious (for instance, in claiming that it is feasible for me to transplant the heart

---

[4] Another reason why the CPA fails as a guide for feasibility regarding complex social or global outcomes.

[5] A similar solution is introduced by Guillery (*2021*), who adds to Wiens's account that the agent needs to be able to bring about the outcome "competently" and "safely." This means that "in all the sufficiently close possible worlds to w (the possible world in which X brings about O), in which circumstances are relevantly similar, X succeeds in bringing about O" (*Guillery 2021, 504*). However, the question remains why stop there. It is still an open question what would be "sufficiently close" to be safe and thus count as feasible, in contrast to cases that are counterfactually fluky and thus infeasible. I doubt that an analysis of these terms can be given that circumvents the need for value judgments.

of a patient) than in instances where less is on the line (for instance, when I say it is feasible for me to return the book tomorrow).

Understood this way, we face a classic case of inductive risk (cf. *Douglas 2000, 2009*; *Wilholt 2009*). Inductive risk occurs when making judgments involves a risk of error that likely has practical consequences. The classic case, described by Rudner (*1953*), is when scientists need to decide whether to accept a hypothesis based on the available evidence. As we rely on scientific knowledge in many ways, it being wrong can have practical consequences that must be evaluated in the scientific process (see Sect. 6.8 for a more extensive discussion of inductive risk.) In some cases, the consequences of false positives, that is, wrongly accepting a judgment, are higher than the consequences of false negatives, that is, wrongly rejecting judgments. For example, if a mechanic needs to judge whether a race car is safe to drive, falsely judging it to be safe would likely have more severe practical consequences than saying it is unsafe and being wrong. Thus, she should be relatively sure to make that call. In other cases, the stakes are less high. For instance, if I know the movie starts late anyway, it is okay to think I will be on time, even if the evidence is not solid.

As one needs to deal with uncertainty in making feasibility claims, and since making judgments under uncertainty in a practically relevant context gives rise to inductive risk, this marks a value dimension of feasibility judgments. This addresses the *problem of defining a threshold* head-on, as I introduced it above. I argued that the common conceptions of feasibility encounter this problem. In the analysis of Brennan and Southwood (*2007*), a feasible outcome is one that has a reasonable chance of being realized by us, given that we try. It must define what "reasonable" means.

The Restricted Possibility Account (RPA) aimed to circumvent this problem but was vulnerable to counterexamples involving "counterfactual flukes." Recall that I described how Wiens's account is vulnerable to being too permissive in instances, where there is an accessible outcome given our resources and causal and conversion processes, but which we would be inclined not to call feasible (cf. *Southwood 2018, 4–5*). Such outcomes are practically possible but involve instances of brute luck, like the example above of buying a private plane by first winning a lottery. This might be a reason for Wiens to restrict his conception to state only necessary but not sufficient conditions of feasibility.[6] Thick feasibility, as I introduce it, allows us to address this problem. This determination of a threshold is less puzzling if we accept that it is partly a value decision to consider an outcome certain and reliable enough to count as feasible. While there

---

[6] Cf. *Wiens 2015, Footnote 1.*

is no general rule for defining a threshold, any statement on the feasibility of an outcome will involve some value-laden evaluation of uncertainty.

This is more than a theoretical exercise. Take the example "LÜTZI BLEIBT" again. Of course, we will want to disallow instances of sheer luck to count towards assessing the feasibility of keeping warming below 1.5 °C. For instance, there might be a magical change in the thinking of all other nations, which makes the burning of coal in Lützerath suddenly compatible with 1.5 °C, because it frees up carbon budget. But this small chance should not count towards its feasibility. Similarly, claiming that Berlin can be climate neutral by 2030 cannot depend on unexpectedly fortunate circumstances, such as that there will be some sci-fi invention that makes it possible. We need to be realistic. However, where to draw the line is tricky, and any answer to it will involve value judgments.

Extending Wiens RPA, thick feasibility is then the following:

Thick Restricted Possibility Account: An outcome is feasible if there is a resource stock that enables us realize it and that is reliably attainable without relying on unacceptable means or producing unacceptable side effects.

If we accept the thickness of feasibility, discussions on feasibility anticipate specific value questions. However, value judgments are only legitimate in limited roles. Thick feasibility includes a value-laden preselection of acceptable means and the determination of an acceptable level of side effects and uncertainty. As we will see, actual feasibility claims do not always restrain themselves to these limited roles. The following section will go back to the normative roles of feasibility judgments in light of the term's thickness.

3.5 Normative Role and the Asymmetry in Making Feasibility Judgments

Thick concepts license "normative inferences," E. Anderson (*2002*) writes. Judgments involving thick concepts can thus make certain normative conclusions without relying on a normative premise. This section explains the normative role of feasibility. It also answers the second open question of the RPA, answering how feasibility can sometimes guide us.

As described in Chap. 2, feasibility is commonly taken to be normatively consequential. Two roles of feasibility stand out: ruling out proposals and giving comparative evaluation in terms of feasibility. The most common role of feasibility is that *infeasible* proposals can be ruled out. There is an extensive debate on how this is to be understood. While a narrow understanding of feasibility as only logical and physical impossibility would suggest a role closer to ought-implies-can (cf. *Gilabert and Lawford-Smith 2012*; *Wiens 2015*), this cannot be the normative inferences of thick feasibility claims. As thick feasibility claims make substantive value judgments, it seems unwarranted to rule out proposals without making the ethical reasons explicit.

The role that fits better with the conception here is what Southwood (*2022*) calls feasibility as "deliberation-worthiness." To call something feasible is to claim it is worthy of further analysis. As infeasible outcomes in the thick concep-tion are either impossible for purely empirical reasons or involve *unacceptable* means, side-effects, or reliance on luck, such outcomes are judged to be unrea-sonable to consider further. All thickly feasible options, however, are part of the set of options that should be considered more fully. Therefore, feasibility judg-ments are about "whether there is a decision to make" regarding some option, not whether to do that action (*Southwood 2022, 131*). To call 1.5 °C infeasible thus would amount to dismissing that we need to think about it any further. If something is feasible, it is worthy of deliberation concerning what we should do. This is compatible with thick feasibility, as I outlined above. Feasibility as a *thickly* viable path means that there is a trajectory that crosses a threshold of being acceptable, which licenses further deliberation.[7]

However, importantly, thick feasibility does not license embarking on the path or pursuing the outcome. There is no way how binary feasibility gives us a reason to pursue a particular outcome or action. Just because some outcome is feasible does not give us a reason to realize it. There is something like an *inductive asymmetry* of feasibility: while feasibility claims have a low positive inductive potential, they have a high negative inductive effect. To state that some outcome or pathway is feasible tells us very little of substance. We cannot infer from such claims any guidance on what to do. We cannot infer anything but that the goal and pathway are worthy of deliberation. They are an option. Whether to take that option is a different matter. In contrast, claims of infeasibility have arguably *high negative inductive effects*. To dismiss an outcome, which is in itself desirable, might have serious consequences. If we follow the normative role of

---

[7] Note that Southwood dismisses a thick reading of feasibility since unacceptable costs are relevant to whether we "ought" to deliberate on an option but not whether it would be fitting to deliberate on an option (*Southwood 2022, 144*). While Southwood is after a philosophical definition, I am looking for a conception for guiding the scientific advising context. I take it that feasibility comes with some normative constraints as well. Thus, in this thick sense, feasibility claims also exclude options that Southwood calls "morally heinous" (*Southwood 2022, 131*).

the judgment, this option is excluded from any further practical thinking and evaluation. This makes it relatively unlikely that we will ever realize it in practice. If it is highly desirable, the practical stakes of calling an outcome infeasible are high. I will return to this asymmetry in Chap. 5.

These discussions, though, bring up the question of how feasibility assessments can provide guidance that many ascribe to it. This was the second challenge to the RPA. Recount that Gilabert and Lawford-Smith (*2012*) called binary feasibility a rather "blunt tool," and therefore, introduce a scalar reading of feasibility. Others agree. For instance, Fabian Schuppert writes that "[v]irtually all statements of what is economically, socially, or politically feasible fall into this [the scalar] category" (*Schuppert 2021, 157*). The CPA received wide praise for relieving feasibility from an unduly and practically irrelevant conception of feasibility as pure possibility. Many take feasibility assessments to have some practical relevance, but the gatekeeping role for deliberation seems to block this.

I take a related sense of feasibility judgments, which I do not count as feasibility claims in the narrow sense, to allow for further normative inferences. This broader sense is value-laden in ways that do not confer to the narrow roles outlined above. This sense is referred to when an outcome is described as "less feasible," "involving greater challenge," or "less realistic." The scalar sense of feasibility introduced by Gilabert and Lawford-Smith (*2012*) is also widely appreciated in publications in the social sciences. Brutschin et al. (*2021*) base their framework entirely on this scalar sense. Other IAM publications similarly rely on a scalar sense of feasibility, for instance, when Gambhir et al. (*2017*) show that the 2 °C goal is "much more challenging [...] when compared to the 2.5–4 C goals, across virtually all measures of feasibility." They also speak directly of the 2 °C goal being "much less feasible" (*Gambhir et al. 2017, 1*). Others use similar terms, such as involving more "feasibility concerns" (*van de Ven et al. 2023*) or being more or less realistic. Clearly, such propositions are reasonable and have something to do with feasibility. Any conceptual discussion on feasibility should be able to account for them.

However, we should avoid jumping toward a scalar conception of feasibility. *First* of all, switching to terms like "realistic" or "challenging" indicates a slight change of topic. For instance, concepts such as challenging and realistic are no longer derivable from the core meaning of feasibility as accessibility. Gilabert and Lawford-Smith (*2012*) introduced the rough meaning of feasibility of an outcome as about whether "there is a way we can bring it about" (*Gilabert and Lawford-Smith 2012, 809*). It is unclear how a scalar notion of feasibility can make sense of this, as there is not such a thing as "more of a way" to bring something about. Gilabert and Lawford-Smith (*2012*) solve this issue by shifting the meaning of feasibility towards a probabilistic understanding, which more naturally lends itself to a scalar reading. We should, though, reject this probabilistic conception (at least in the context of feasibility assessments); at least, this is what I argued in the last chapter.

However, we might think of the scalar use of feasibility as a *derived sense* of feasibility, which involves further value judgments concerning the selection and weighing of different aspects of trajectories. Take the case of Gambhir et al. (*2017*), who analyze some pathways to be more challenging than others. In comparing the "feasibility," the authors not only have to weigh the economic costs of an option to other considerations, such as the expected social resistance or the effect of alienating some allies, but they also have to give some account of what should count as costs. We must make value assumptions to determine such criteria in terms of feasibility. Claiming an option or goal is "more feasible" than alternatives will involve various value judgments.

If there is no scalar sense of feasibility itself, such claims on comparative feasibility in terms of the degree of challenge, realism, or concern should be taken more broadly as a form of *evaluative* reasoning based on feasibility assessment. What can this look like? Strictly speaking, finding one pathway to show that some state of affairs is (narrowly speaking) feasible would be sufficient. However, this would be poor scientific advice to policymakers concerning the feasibility issue. Policymakers need to know not only that a state of affairs is feasible but also how it is feasible and what different path leads to its realization. As scientists should not settle policy decisions, feasibility assessments based on the restricted possibility account should aim to explore a wide range of viable pathways.

Such an exploration will reveal all kinds of facts relevant to the comparative evaluation of pathways. Such normatively relevant feasibility facts consist not only of the necessary steps for realizing a desirable institutional scheme but also concern side effects, implications for other goals, dependencies between different actions, the kind of agents that need to be involved, the costs associated with different pathways, etc. Once we combine these "feasibility facts" with normative considerations, we derive comparative evaluative statements about the pathways in question (cf. *Brennan 2013, 316*). Answering the question of what is more challenging or realistic will depend on evaluating these implications, aggregating costs of diverse kinds within different pathways, and aggregating these normative considerations reasonably. Any of the steps involved will depend on value judgments beyond the narrow roles outlined in the thick conception above. There are no relevant propositions in such an understanding, which helps to compare proposals without making a value judgment.

Thus, the challenge of *comparative evaluations* can be met, though only in a way that is possible by engaging more fully with value questions. Such statements on some pathways being more challenging or less feasible depend on implicit or explicit value assumptions. If one follows these normative assumptions, such derivative feasibility claims can be action-guiding in a fuller sense than outlined above: they give us a preliminary reason for action. However, according to my account, such claims are not proper feasibility judgments.

Normative inferences based on thick feasibility are thus twofold. Proper, binary feasibility claims warrant excluding options and pathways from deliberation. This implies that judgments concerning the infeasibility of a particular goal and option have significant consequences for our practical deliberation and, thus, must be well-founded. This sense only allows specific restricted roles for value judgments. In a second, derived sense and role, feasibility claims involve more value judgments and license further normative inferences. If an option is less challenging, this might give us a reason for action. However, we need to specify what this term means in such cases, as it deviates from the meaning of feasibility.

3.6 Summary

This chapter defended a value dimension of feasibility. I started by outlining the widely held presumption of *Descriptive Feasibility*, which takes feasibility to be an empirical term free of normative content. *Descriptive Feasibility* would allow for clearly separating descriptive feasibility facts and aspects pertaining to values.

Some philosophers have argued, in contrast, that feasibility involves normative aspects, and I surveyed their arguments. As it is discursively used, I argue that feasibility involves value judgments and that this value dimension can help solve the two open questions the RPA left us with. Thick feasibility explains how feasibility can be more substantive than the merely practically possible, as it involves additional normative considerations and helps define a threshold between the possible and the feasible. Moreover, by allowing additional value judgments in our assessments, we can make comparative evaluations, which in a derived sense have to do with feasibility.

At its core, however, feasibility is thick in three specific ways: feasibility judgments involve normative assumptions concerning what means and concerning what side effects are acceptable. Further, normative evaluations are necessary to determine what level of uncertainty can be tolerated when claiming an outcome to be feasible.

Finally, I specified the normative role of feasibility as a gatekeeper for deliberation. If an outcome or proposal is infeasible, we need to consider it no further in our practical deliberation. This normative role makes feasibility claims asymmetric in the sense that negative claims of infeasibility are more consequential than positive claims. That an outcome or path is feasible gives us no reason to pursue it. It simply means we can and should deliberate upon it.

Part II
Modeling Feasibility

Integrated Assessment Models

4

Integrated Assessment Models (IAMs) have become one of the central tools of scientific advice to policymaking in the climate context. Despite being a relatively small field, contributions from IAMs, for instance, make up a fifth of all publications in the IPCC Synthesis Reports (*van Beek et al. 2020, 2*). However, IAMs have also seen extensive criticism, primarily concerning their large-scale reliance on Carbon Dioxide Removal (cf. *S. Beck and Mahony 2018b*) and their lack of transparency (cf. *Robertson 2021*). This chapter retells the history of the models, delineates the kind of models I am concerned with, explains their working, and provides reasons why we should care about their results.

The chapter first gives a short history of IAMs (Sect. 4.1), distinguishes Process-based IAMs from the other types of IAM (Sect. 4.2), briefly explains the inner makeup of PB-IAMs (Sect. 4.3), outlines a general case for why we need to rely on IAMs (Sect. 4.4), describes how IAMs use scenarios (Sect. 4.5), and finally explain how the question of feasibility has become a central research question of IAMs (Sect. 4.6).

4.1 A Short History of Integrated Modeling

In 2023, CO_2 concentrations stood at 419 ppm (*NOAA 2024*), higher than any point in at least 800,000 years, and concentrations continue to rise yearly. IAMs are a central tool for understanding the underlying dynamics behind this continuous rise in atmospheric concentrations of greenhouse gases and, even more importantly, a tool to understand how to reverse this trend. This section presents a short history of IAMs

© The Author(s) 2025
S. Hollnaicher, *Assessing Feasibility with Value-laden Models*,
https://doi.org/10.1007/978-3-662-70714-2_4

The first assessment reports concerning the "CO_2 problem" were published back in 1979 with the Charney report (*Charney and et al. 1979*) and the 1983 National Academy of Science report "Changing Climate" (*Research Council 1983*), which linked possible climate impacts to the continuing burning of fossil fuels and deforestation. Integrated Modeling started even a few years earlier when one of the first global modeling exercises was performed with a model called *World3*. This model served as the basis for the famous 1972 report "Limits to Growth" report (*Meadows et al. 1972*). World3 did not represent the climate system, nor was the report concerned with climate change. It was, though, the first influential socio-economic computer modeling exercise that quantified and problematized human action on a global scale, delineating our planetary boundaries. As it integrates different systems within a modeling framework, we may consider it to be the first IAM (*van Beek et al. 2020, 4*). Curiously, much of the controversy surrounding methodology and modeling assumptions, which is part of the IAM discourse today, was raised against World3 (*Blanchard 2010*).[1] However, the particular framework of World3, focusing on the limits of growth, was abandoned and made way for a different (growth-based) approach, which has achieved niche hegemony today and makes up much of IAM discourse (cf. *Purvis 2021*).

The publications by William Nordhaus established a different understanding of the climate change problem in the 1970 s than the one World3 was proposing. Nordhaus was inspired by the environmental problem that the Limits to Growth report raises but was also profoundly unsatisfied with the Malthusian thinking it used to tackle the problem. He, thus, took a different approach by using existing energy system models, adding a climate module to it, and establishing the first IAM in the model *DICE* as we think of them today. The central premise of this model is to think of climate change as a balancing problem between the costs of mitigation and climate impacts. DICE is used to mathematically calculate "the efficient path of resource extraction and depletion" that balances these two sides of climate change (*Nordhaus 2019*). DICE provides projections of atmospheric CO_2 concentrations based on different socio-economic assumptions. It calculates so-called "shadow prices" for carbon emissions, which we now know as the "social costs of carbon" (*Nordhaus 2019*). He later (in 2018) received the Nobel Prize

---

[1] Blanchard (*2010*), for example, describe the development of a Latin America Model in response to the implicit "stabilization of the world economy in its present structure, which would maintain North-South inequities. The Bariloche project explicitly attempted to avoid such an outcome" (*Blanchard 2010, 109*). We will see that such implicit inequalities in future energy demand development are still controversial value-laden assumptions in IAMs.

for "integrating climate change into long-run macroeconomic analysis," including the development of DICE.

Nordhaus pioneered the economic significance of climate change and the need for avoiding business-as-usual, warning in 1979 already that "we are probably heading for major climatic changes over the next 200 years if market forces are unchecked" (*Nordhaus 1979*, IIX-IX). Nordhaus' conceptualization of climate change as a balancing problem profoundly influenced mainstream economic thinking on climate change. Nordhaus, moreover, was an active policy advisor and established the significance of IAMs in the science-policymaking interface. The social cost of carbon, provided by the DICE or similar models, still plays a significant role in US policymaking. It also inspired researchers to use complex energy system models and economic welfare models and apply them to the mitigation side of the balancing problem in greater detail.

The 1980 s brought a lot of model development and the establishment of the IPCC, which became a central hub of IAM research. The model *IMAGE*, developed at the IIASA, provided (in a shared effort with the ASF model[2]) high and low-emission scenarios for the First Assessment Report of the IPCC (*IPCC 1990*). Such emission profiles proved versatile and provided a link for research in different fields. For instance, scientists could project the climate impacts of different scenarios by using emission and land-change projections from IAMs combined with more complex climate models. Around this time, the global community also established the UNFCCC process to target the climate problem politically. This gave the field of integrated modeling another push. The UNFCCC process sparked many questions that pure climate models could not answer.

Mitigation pathways from IAMs became the "backbone of scenario analysis of Working Group III of the IPCC" (*van Beek et al. 2020, 2*). In the 1990 s, IAM science hit off. Before 1990, a Google Scholar found only 34 publications on "integrated assessment models." The 1990 s added over 800 publications, and in the 2000 s, another 3.000, with exponentially more following after that. While there were only three IAMs around in 1990, this number grew to forty models in 1997 (*van Beek et al. 2020, 6*). As a leading modeler put it, the most recent history of global climate governance and science advice established "a place in the sun" for IAMs.[3] In the 2010 s, temperature goals became the focus of climate governance, and IAMs were the tools in place to translate these abstract climate goals into concrete mitigation strategies and measures. The modeling of the feasibility of 2 °C as a target was done in 2007 and judged "stabilization

---

[2] The "Atmospheric Stabilization Framework" (ASF) model (cf. *Lashof and Tirpak 1990*).

[3] Taken from a presentation by Detlef van Vuuren at the IAMC 2022.

as low as 450 $ppmCO_{2-eq.}$ to be technically feasible, even given relatively high baseline scenarios" (*van Vuuren et al. 2007, 119*).

Studies such as these established a new framing of climate economics. Economic modeling in this new approach aimed not to balance the costs and benefits of climate change but to take climate goals as a scenario input and find efficient and feasible trajectories for staying within these goals. This target-based modeling increased significantly with the establishment of the Paris goals, as I will describe in more detail below. Between the AR4 and AR5, the ratio of IAM contributions in the IPCC reports (while being only a small fraction of the general scientific discourse on climate mitigation) doubled from 10 to 20 % (*van Beek et al. 2020, 2*). Notable models were added in this period, including the *ReMIND* model developed at the Potsdam Institute for Climate Impact Research (PIK), one of the leading centers of IAM research today (*Luderer et al. 2015*). IAMs have become a central tool in the science-policy space.

The last section of this chapter continues the history of IAMs to its most recent development, as IAMs are increasingly used to assess the feasibility of climate goals and different mitigation strategies. First, though, the following section explains what IAMs are and delineates the kind of models this book investigates.

4.2   Demarcating Process-based IAMs

There are a lot of "climate models" around, and philosophical contributions often do not pay enough attention to their differences.[4] This section specifies the kind of models I will investigate. This book concerns complex *Process-based IAMs*, which take temperature goals as a scenario input and model mitigation pathways compatible with these goals.

In its most general self-characterization, Integrated Assessment Models are models that "aim to provide policy-relevant insights into global environmental change and sustainable development issues by providing a quantitative description of key processes in the human and earth systems and their interactions" (*IAMC 2022*). For this purpose, IAMs link the human system to the earth and climate system. The IPCC defines IAMs as "simplified representations of complex physical and social systems, focusing on the interaction between economy,

---

[4] Intemann (*2015*), for example, targets value questions in Global Circulation Models (GCMs) and value questions in IAMs, including Cost-Benefit Models. GCMs, however, are used to make predictions on future warming. CB-IAMs instead make normative claims about the best climate policy. Thus, the two kinds of models have a vastly different relationship to value questions.

society, and the environment" (*Guivarch et al. 2022, 1843*).[5] In modeling these interactions, IAMs aim to provide projections of the future based on scenarios concerning the main drivers of carbon emissions and significant technological and social opportunities to mitigate climate change.

The three parts of the term describe key characteristics of IAMs. The modeling is *"integrated,"* as it combines knowledge and engages researchers from various disciplines across the natural and social sciences. Nordhaus defines IAMs as "approaches that integrate knowledge from two or more domains into a single framework" (*Nordhaus 2013, 1069*). Most IAMs I discuss involve more than two domains, aiming to provide an encompassing perspective on climate mitigation. IAMs provide *"assessments,"* as they are specifically built and used to investigate policy-relevant research questions and provide informative answers to questions in policymaking. The most important venue of IAM research is the Working Group III report of the IPCC and other Global Environmental Assessments (GEA) reports (cf. *Kowarsch, Jabbour, et al. 2017*), including many national and regional studies on climate mitigation strategies. Finally, IAMs are *"models,"* as they are a numerical and highly abstract representation of the world and the relevant subsystems, expressed in formulas and thousands of lines of computer code.

Underneath this general characterization is a large diversity of models. The most basic distinction is between two major types of IAMs (cf. *Weyant 2017, 117*). One branch of models is *Cost-Benefit IAMs* (CB-IAMs), sometimes also referred to as "policy optimization models" (*Weyant et al. 1996; Nordhaus 2013, 1080*). The three most notable CB-IAMs are the already mentioned DICE ("Dynamic Integrated Climate-Economy model") model, developed by Nordhaus (*2010*), PAGE ("Policy Analysis of the Greenhouse Effect"), by Nicholas Stern and used in the Stern Review, and FUND ("Climate Framework for Uncertainty, Negotiation and Distribution") by Richard Tol. CB-IAMs use highly aggregated representations of the economic and climate system. In DICE, for instance, the climate impact side is approximated by a single formula linking global warming to a level of economic damage in terms of GDP.

The distinguishing feature of CB-IAMs is that they aim to determine an optimal overall level of climate mitigation by balancing the various impacts of climate change with the costs of avoiding them through climate mitigation (*Weyant 2017, 117*). Pathways determined by CB-IAMs represent the "optimal" balance

---

[5] A detailed overview is given in Annex III of the IPCC AR6 (*IPCC 2022, 1843–70*; cf. *Weyant 2017*). An excellent low-level introduction can be found by Evans and Hausfather (*2018*).

between climate impacts (and benefits) and costs (and side-benefits) of mitigation measures based on the scenario assumption.[6] This approach has received much criticism, conceptually and substantively, for the concrete assumptions used. Nordhaus' modeling, for instance, has suggested an optimal warming level of around 3 °C as late as 2017 (cf. *Nordhaus 2017*), a result highly dependent on a few contested assumptions.[7] Many critics have pointed out the ethically questionable assumptions underlying such modeling (cf. *Gardiner 2011*; *Frisch 2017*; *Keen 2021*; *Keen et al. 2021*; *Smith 2021*). However, as this book concerns the second strand of IAMs, I will not engage with these discussions in more detail.

This book investigates *Process-based IAMs* (PB-IAMs), also known as "policy evaluation models" (*Weyant et al. 1996*; *Nordhaus 2013, 1080*) or sometimes simply as "complex" or "large-scale" IAMs (*Clarke et al. 2014, 422*). As the name suggests, these models are typically more detailed in their representation of the various systems. They often include some sectoral and regional segmentation and disaggregated representation of different mitigation options and technologies, which makes them applicable to a range of new research questions. The main feature of PB-IAMs is that they focus only on mitigation. The models do not represent the impact or damage side of climate change. Instead, temperature goals are applied to the model exogenously as a scenario input.[8] The main research question of PB-IAMs thus becomes how specific predefined climate goals can be reached, subject to a set of scenario assumptions concerning constraints and enabling conditions of climate mitigation.

With this general distinction at hand, let us have a closer look at PB-IAMs. (From here on, I use "IAM" to refer to PB-IAMs, unless specified otherwise.) While national and regional IAMs are also essential in advising policymakers, this book focuses on *global* IAMs. The IAMC wiki documents a total of 27 global IAMs. The IPCC AR6 considered scenarios from 50 different models (*Riahi et al. 2022, 306*). However, despite this diversity, a few models and research

---

[6] Another significant output of such models is the social costs of carbon (SCC), which vary greatly depending on key (value) assumptions but play a highly influential role in US policymaking (*Frisch 2017*; *Backman, Burke, and Goulder 7.6.2021*).

[7] An updated representation of the climate impact side, for instance, has challenged Nordhaus' conclusions and argued that the optimal level, even within his approach, is closer to 2 °C (*Hänsel et al. 2020*).

[8] *"Endogenous"* variables are modified internally by the model, while *"exogenous"* parameters and variables are applied externally to the models. Temperature goals (or corresponding carbon budgets or RCPs) are provided to the models as explicit input and are, thus, exogenous.

hubs provide a large proportion of mitigation pathways. In the AR6, of the 1,686 scenarios considered, one-third came from the two most significant contributors:

- **ReMIND** ("Regional Model of Investment and Development"), developed by the Potsdam Institute for Climate Impact Research (PIK) (*Luderer et al. 2020*).
- **MESSAGE** ("Model for Energy Supply Strategy Alternatives and their General Environmental Impact"), developed by the International Institute for Applied Systems Analysis (IIASA) (*Krey et al. 2020*) in Laxenburg, Austria.

The following four models produced another fourth of the 1.686 scenarios considered in the AR6:

- **IMAGE** ("Integrated Model to Assess the Global Environment"), developed at the PBL Netherlands Environmental Assessment Agency in the Netherlands (*Stehfest et al. 2021*; *Roelfsema et al. 2022*)
- **POLES** ("Prospective Outlook on Long-term Energy Systems") developed in a joint EU Science Hub and used, for instance, by the European Commission in its "Global Energy and Climate Outlook" (*Després et al. 2017*);
- **WITCH** ("World Induced Technical Change Hybrid") developed at the RFF-CMCC-EIEE European Institute on Economics and the Environment in Milan, Italy (*Bosetti, Massetti, and Tavoni 2007*; *RFF-CMCC-EIEE 2023*).
- **GCAM** ("Global Change Analysis Model") developed and used by the Joint Global Change Research Institute (JGCRI) at the University of Maryland and the Pacific Northwest National Laboratory (PNNL) (*Calvin et al. 2019*).

These six models make up more than half of the scenarios in the AR6. The research teams and the institutes behind them also play a leading role in coordinating the research more generally. Most of the members of the *Scientific Working Groups* of the IAMC[9] and all but two co-chairs[10] come from these institutes. Four of the five *Scientific Working Groups* are co-chaired by a PIK or IIASA researcher. IIASA hosts the AR6 scenario database (*Byers et al. 2022*), coordinates the project that provides almost half of the scenarios for the AR6 (the ENGAGE project,

---

[9] The *Integrated Assessment Modeling Consortium* (IAMC) is a joint organization of research institutions that engage in Integrated Assessment Modeling and analysis. It was created in 2007 after the IPCC called for an independent body to lead the provision of emissions scenarios to keep the IPCC as independent assessors of research who do not conduct research themselves.

[10] The other two are Shinichiro Fujimori of Kyoto University and Roberto Schaeffer from the Universidade Federal do Rio de Janeiro.

Riahi et al. 2021), and provides the coordinating lead author for the IAM-focused Chap. 3 of the AR6. Thus, much of scenario production happens within a few research hubs. All six main contributors are in the USA or Europe. While some members are from other parts of the world, the only notable contributor to the scenario evidence from the Global South is the COFFEE-TEA model, developed at COPPE/UFRJ in Brazil.[11] The yearly conference of the IAMC in 2023 had one participant from Africa (participating online). It is fair to say that IAM science happens mostly within Western institutions.

PB-IAMs are the focus of this book, and I will explain them in more detail in the following section. IAMs are computer models that combine representations of the socioeconomic system with the climate system to model possible trajectories for the future. The distinguishing feature of PB-IAMs is that they take tempera-ture goals as an explicit input and compute feasible and cost-efficient pathways to stay with these goals based on scenario assumptions. While there are various IAMs, a handful of models from Europe and North America comprise the larger part of scenario data from IAMs.

4.3   A look into Process-based IAMs

If one is not already dizzy from all the abbreviations in the last section, a sus-tained look at Fig. 4.1 might do the job. The figure gives an overview of the IMAGE model, one of the Top 2 contributors to IAM pathways. It exemplifies the complexity of variables, modules, and interactions considered within a model, displaying only the most critical of them. This section will go through some parts of the model in more detail to better understand how IAMs function. IAMs typi-cally at least involve a representation of the economy, the energy system, the land system, and the climate system, so I will quickly go through these four. Keep in mind that different IAMs vary in the way they operate.

---

[11] The "Computable Framework For Energy and the Environment-Total Economy Assess-ment," based on MESSAGE by the Coimbra Institute for Graduate Studies and Research in Engineering of the Universidade Federal do Rio Janeiro.

Image 3.0 in detail

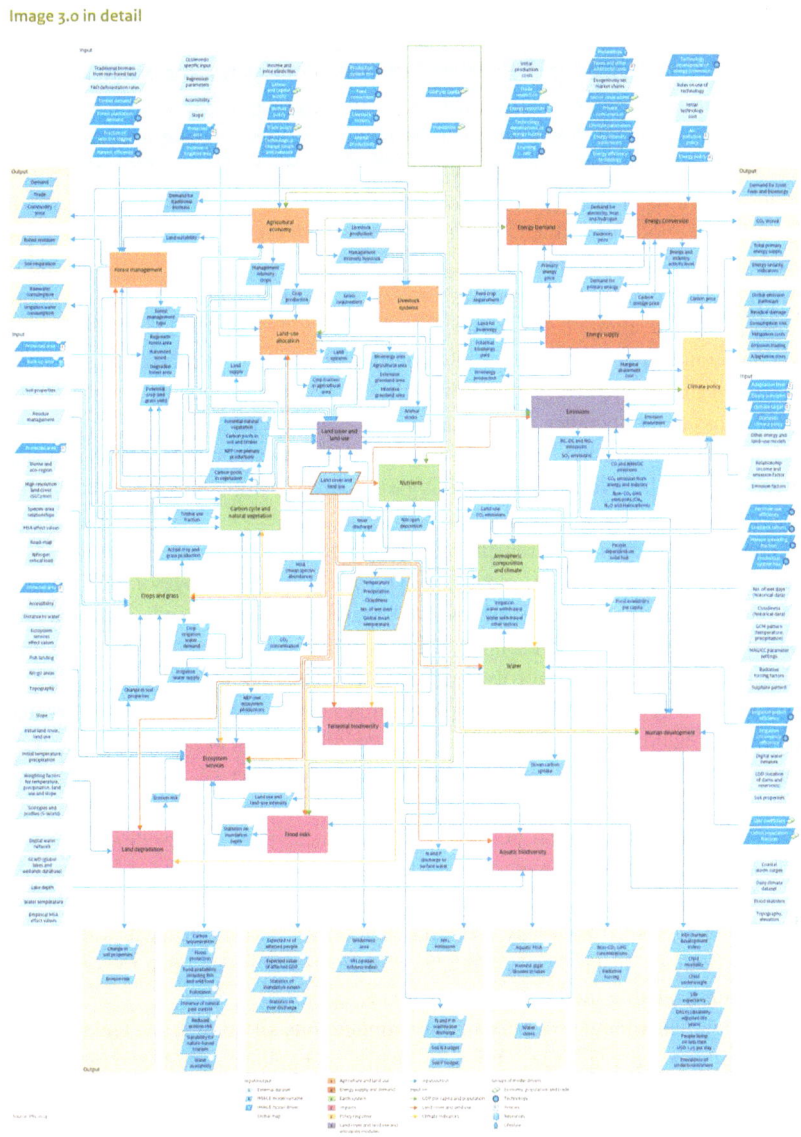

Fig. 4.1 Overview of the IMAGE model, URL: https://models.pbl.nl/image/Big_Flowchart cf. Stehfest (2014)

The *economic system* is typically the core of IAMs, coordinating the various interactions between sectors to derive a cost-efficient pathway (at least in macroeconomic growth models).[12] Such macroeconomic modules, for example, in the ReMIND model, operate with variables such as capital stock, final energy, and available labor. These variables (with the help of a range of parameters) are used to determine the total economic output. Representative households are used to model consumer decisions and determine the demand for different energy services. These modules influence energy demand, which interacts with the supply side systems, for example, the prices determined in the energy modules. Many IAMs, such as the ReMIND and MESSAGE, come from a background in welfare economics. The internal target function then maximizes welfare approximated by total consumption. I will get into the details of these value-laden modeling choices in more detail. In general, though, the economics modules bind the different subsystems together to give an overarching macro perspective on climate mitigation. They further produce some of the critical outputs of IAM pathways, such as the overall costs of mitigation strategies or their macroeconomic implications.

The *energy system* is often the most detailed part of IAMs, as IAMs often evolved out of energy system models. It typically includes a range of technologies for providing end-use energy demand. Demand is a function of other macroeconomic variables, e.g., household consumption and industry output. Efficiency gains in energy production or use are explicitly modeled as a response option or assumed as a set of fixed internal parameters. The energy module ensures that end-user demand is always met using the available energy resources and conversion technologies. Each scenario run must define the set of energy parameters (e.g., cost, potential, emission factors) and constraints to the energy model (e.g., sociopolitical constraints on possible build-up rates). The energy model ensures that end-user demand is met in a cost-efficient way. In doing so, the energy module interacts in various ways with other modules. Bioenergy, for example, is valuable in IAMs across many different sectors. As it competes with other land uses, the reliance of the energy module on bioenergy is limited due to constraints concerning available land and ecological side effects.

Land system modules represent the implications of mitigation technologies regarding land and water use. One land-intense technology included in most IAM

---

[12] Partial equilibrium models only represent a subset of markets and do not model all interactions of the economy (*Guivarch et al. 2022, 1845*). Another set of models is General-Equilibrium Models, or Computable General Equilibrium (CGE), which "represents the economic interdependencies between multiple sectors and agents, and the interaction between supply and demand on multiple markets" (*Guivarch et al. 2022, 1845*).

pathways is bioenergy, including its use in BECCS. BECCS, *BioEnergy with Carbon Capture and Storage*, has played a controversial role in recent pathways and will, thus, come up in multiple discussions throughout this book. In BECCS, photosynthetic processes capture atmospheric CO_2 in biomass, which is subsequently utilized in various processes for energy use or biofuels. If one captures and stores the carbon released in these processes, CO_2 is permanently removed from the atmosphere, and one has achieved so-called "negative emissions." As bioenergy depends on cropland availability, it affects food production and other competing land usages. To model such interactions, gridded representations of the land systems and their characteristics are included in the models (*Guivarch et al. 2022, 1857*). The ReMIND model is for this purpose used in combination with the land model MAgPIE, "a global land use allocation model" that represents the land surface on a rough grid, computing "specific land use patterns, yields and total costs of agricultural production" based on regionalized bioenergy and food demand (*IAMC 2021*). Food demand is calculated from population numbers based on the (economically adjusted) local diet. The land module further calculates the emissions from other land-based activities, such as emissions from agricultural activities or changes in forest cover.

Finally, the task of the *climate module* is to model the climate forcing resulting from the various activities determined by the other modules. This module translates energy production, transportation, land-use, and industry processes into greenhouse gas emissions and resulting global temperature trajectories. MAGICC, a simplified version of a more complex climate model, commonly does this. It is highly reduced in complexity but provides a fair approximation of the current scientific understanding of the climate response to human action. As MAGICC is further highly efficient concerning computing time, it gets widely used in IAMs.

This quick tour through the inner workings of PB-IAMs gives an overview of the most important modules and interactions. Concrete IAMs are way more complex, as every process involves numerous variables and routines. While IAMs have different strengths and weaknesses, all provide a relatively coarse representation of any single sector or process. The strength of IAMs is that they model all sectors and interactions at the same time. This overarching perspective is the base for the case for IAMs, outlined in the next section.

4.4   The Case for IAMs

It is fair to say that Integrated Assessment Models have been immensely successful in the science-policy interface in recent years. Take, for instance, the two quotes collected by van Beek et al. (*2022*), which highlight the exceptional role IAMs play:

> "'If we did not have IAMs, we'd have to invent them because they are the only way of getting between human activity on climatic changes on a century scale' (interview 2, IPCC Bureau member)" (*van Beek et al. 2022, 197*).

> "'Even when I am critical of IAMs and throw them all out of the window, if I sit tomorrow at my desk, I would still build a new IAM. One that understands how decisions in land use or building affect how much mitigation we need and how much land we need' (interview 4, CLA IPCC SR1.5, IAM modeler)" (*van Beek et al. 2022, 197*).

These experts describe IAMs as essential scientific tools for advising on the issue of climate governance. This section elaborates on why this is the case. It presents *the case in favor of IAMs*.[13]

The most general argument for IAMs is the following. Climate change is a global, intertemporal problem caused by greenhouse gas emissions and land use changes. These causes of climate change are closely entangled with everything we do and care for. On one level, of course, the solution is simple—stop burning fossil fuel and changing land. However, taken at face value, this is an untenable proposal.[14] While it might be physically possible to stop all emissions in an instance, the consequences would be devastating. The challenge is to find feasible and just trajectories from where we are towards climate neutrality, which keep much of what we care about alive and fulfill other valuable global goals, such as reducing Global poverty. The dynamics underlying these issues are incredibly complex, as any solution to climate change involves a multiplicity of trade-offs, side-effects, and co-benefits and has costs that evoke intergenerational and global questions of fairness in itself. Scientific advice is of great value in governing this complexity. Since climate change touches on all action in all sectors of human action, one needs an overarching perspective. Nordhaus writes:

---

[13] IAMs did not become highly successful solely because of the general value outlined in this section. Some background conditions helped them achieve success. The rise in computer power, a general "trust in numbers," and cultures of prediction helped make advice from IAMs as influential as it is today (*van Beek et al. 2020, 10*).

[14] I am unaware of anybody defending it, though Tank (*2022*) might come closest in arguing that any luxury emission is morally impermissible as it produces unjustifiable harm.

"The point emphasized in IAMs is that we need to have *at a first level of approximation* models that operate all the modules [carbon cycle, climate system, climate impact, climate policy] simultaneously" (*Nordhaus 2013, 1976*). IAMs promise to provide this birds-eye view from above.

IAMs help to "give policymakers at all levels of government and industry an idea of the stakes involved in deciding whether or not to implement various policies" (*Weyant 2017, 116*). Moreover, IAMs provide conceptual contributions "for developing insights about highly complex, nonlinear, dynamic, and uncertain systems" (*Weyant 2017, 131*). Determining the stakes and consequences of policy decisions is where economics comes in, more precisely, the subfield of welfare economics, as it is the base for most IAMs. On a theory level, welfare economic theory imagines a central planner who can determine these various trade-offs and relative values by quantifying the benefits and side-effects of different measures. The underlying goal is to maximize society's welfare. It is precisely this quantification that allows for the simultaneous evaluation of very different kinds of goods. Broome (*2012*) writes:

> "Quantitative judgments on this scale demand the methods of economists. Economists are the experts in large and complex aggregations of this sort. They have the mathematical and statistical techniques for making them. In the case of climate change, they have been making them for decades. We depend on their work. In the end, we shall have to rely on the conclusions economists arrive at, because we non-economists cannot do the calculations ourselves" (*Broome 2012, 103*).

According to Broome, the government has, to a large extent, duties of goodness. Such duties do not demand specific actions but require agents to produce the most good concerning various goals with the means available. As there are always competing ways to do good, weighing them against each other becomes necessary.[15] If we follow Broome, policymaking involves difficult trade-offs between different goals and measures, and quantifying the various implications is vital for deciding what to do.

Broome, however, does not think we should trust economists unquestioningly:

---

[15] The counterpart to "duties of goodness" are "duties of justice," which are negative duties that prohibit specific actions. For individuals, "reducing emissions is [generally] a duty of justice and also a duty of goodness" (*Broome 2012, 53*). For governments, however, reducing emissions is only a duty of goodness, which competes with other societal goals and thus needs such analysis. Broome thinks governments sometimes have negative duties, for instance, not to torture. However, he defends that no negative duties related to emission-related activities exist for governments due to the Non-Identity Problem (cf. *Parfit 1984*; *Roberts 2023*).

"[A]s non-economists we can assess the foundations of their work. We do not automatically have to accept the ethical premises that economists themselves assume. [...] Once the correct ethical premises are in place, the methods of economics can be applied to work out their implications. My role as a 'moral philosopher is with the premises rather than directly with the conclusions. I aim to set out a correct theory of goodness, for the methods of economics to put into application" (*Broome 2012, 103*).

As we will see in this book, we need to discuss the ethical assumptions of IAMs for the models to deserve trust in their results. Following Broome, however, we need the analysis economists and modelers provide.

We can see the concrete value of IAMs by looking at a few applications of the models. One concrete application of IAMs is to compare pathways compatible with climate goals with projections for the future based on the actual policies in place. Such an analysis reveals the ambition gap of climate policy. Such analysis is necessary for us to know where we stand and what is necessary to do. Fig. 4.2 is taken from the Summary for Policymakers of the IPCC report. It summarizes the collective political failure in addressing climate change in a single graph. Arguably, this is one of the most significant graphs out there. Even now, in 2023, actual policies lead us sideways in global emissions, while any pathway compatible with a safe planet would point sharply downwards. Such a graph plays an essential role in holding governments accountable (cf. *UNEP 2022*). IAMs are crucial for this task. We need to know what path we would need to take to estimate the failure of our governments.

We also need to know what paths we can take to achieve better futures. It is, as described above, by no means easy to know what to do to stop burning fossil fuels without collateral damage. IAMs provide, on a high level of abstraction, feasible pathways that inform us of ways to achieve a future within the planetary boundaries. As IAMs provide many different pathways, they can provide the ground for discussing what path we should take and which future we end up choosing. In other words, IAMs are valuable tools for assessing feasibility. However, what this exactly means is the topic of the next chapter.

The most general case for IAMs is that they are a tool to provide a quantified way to deal with the interactions between different sectors involved in climate mitigation policy. IAMs are valuable in holding governments accountable, providing insights into different mitigation strategies, and shedding light on what the abstract climate goals demand in practice. However, as Broome and others have

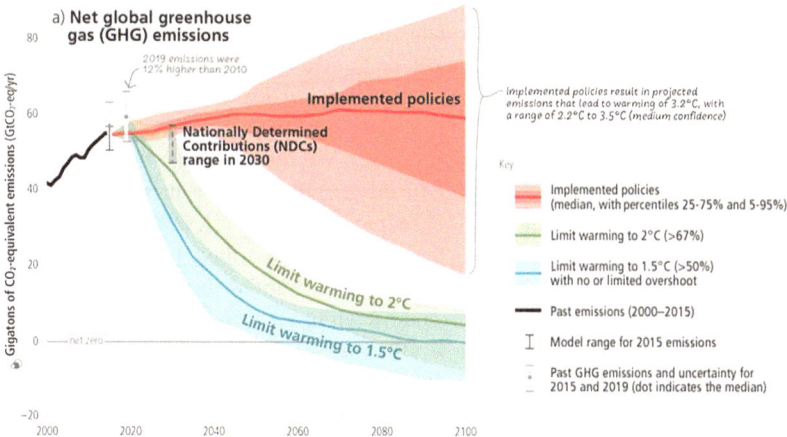

Fig. 4.2 Excerpt from a figure in the Summary for Policymakers (IPCC 2023, fig. SPM.5) "Global emissions pathways consistent with implemented policies and mitigation strategies"

pointed out, we should be skeptical about their results.[16] This book critically engages with IAMs and their ethical assumptions.

Ultimately, IAM results matter since they fill a vital knowledge gap. We need to translate abstract goals into different pathways. Policymakers demand such knowledge on solutions, and IAMs' quantified approach has become a vital element of the climate discourse.

4.5 IAM's Use of Scenarios

This section describes what kind of knowledge IAMs provide. We need to understand models in light of their purpose. The purpose of IAMs is to provide orientation knowledge on climate mitigation strategies to policymakers and the public. IAMs abstract and idealize from reality to provide helpful guidance, which

---

[16] Some philosophers have argued that classical economic theory proved inadequate concerning the climate problem (*Gardiner 2011*). Climate change is a "problem from hell" for economists (*Weitzman 2014*). We should be skeptical of some of the concrete quantifications made in climate economics. This is the task of this book. In the end, I will argue that IAMs should be seen as one value-laden tool in assessing feasibility, and other methodologies must also be considered. However, this does not undermine the general case for IAMs, as presented in this section.

modelers express in a self-characterization as being the mapmaker for climate governance. This section further explains that IAMs rely on a scenario approach.

No philosophical book on models can be printed without quoting Box's dictum that "all models are wrong, but some are useful" (*Box and Draper 1987*). Put another way, models are not correct or false, but they "lie on a continuum of usefulness" (*Barlas and Carpenter 1990, 157*). We must evaluate the models and understand their results with respect to the specific purpose they are built for. The overarching purpose of IAMs is to provide policy-relevant knowledge on ways to mitigate climate change. As the last section described, IAMs are valuable because we need a perspective informing us on the various interactions and trade-offs between different strategies to combat climate change. We need to know the implications of our choices, how effective specific measures are, what co-benefits and side-effects we can expect, and how this adds up in light of the larger goal of staying within safe planetary boundaries. IAMs provide this perspective. The IPCC writes: "key purpose of IAMs is to provide orientation knowledge for the deliberation of future climate action strategies by policymakers, civil society and the private sector" (*Guivarch et al. 2022, 1859*).[17] There are multiple users of IAM scenarios. Activists can rely on pathways to ground their demands for the future in authoritative knowledge. Industry agents can use knowledge from IAMs in their planning as pathways describe different ways a green transition can occur. Most notably, policymakers and the public can rely on pathways to understand what the Paris Goals imply and to help make decisions concerning different instruments and mitigation strategies.

A common image of modelers for their task in the science-policy space is that of mapmakers. In this image, IAMs provide maps for the future, with policymakers being the navigators who have to decide on a course of action (*Edenhofer and Minx 2014*). The map metaphor is a helpful image for various reasons (and far from an uncommon metaphor for science, cf. *Kitcher 2001*). The first aspect of maps is that they provide orientation knowledge. Maps are useful to various users by providing essential knowledge for deciding what to do. Without a map, finding a way is, in many situations, close to impossible. IAMs, in this sense, provide orientation in the "largely unknown territory of climate policy" (*Edenhofer and Minx 2014, 37*). A second aspect of maps is that, as a metaphor, they convey an aspiration. Maps promise to be usable for entirely different purposes. The same map might be used to take a pleasant Sunday stroll through the mountains and to

---

[17] Weyant (*2017*) describes three uses of PB-IAMs: mitigation analysis, impact analysis, and integrated mitigation and impact analysis. The first is the most developed use of PB-IAMs and the focus of this book.

combat a war. The map metaphor communicates the aspiration not to dictate to the users what to do. These goals of neutrality and objectivity will concern us in Part III of this book, which engages with value judgments arising in IAMs and offers strategies to deal with them in a legitimate way.

A third aspect is that maps idealize and abstract from reality in light of their purpose. A hiking map shows only part of the landscape and uses symbols and conventions that are helpful in providing orientation for hikers. Such a map will convey a different representation of the land than a military or geological map. To be of help, Maps need to abstract from reality and idealize features of the landscape. Maps are wrong in the sense that Box said models are wrong. Consider Carroll's famous story of the people who wanted to perfect their map, eliminating all its errors. In the end, they produce a map on the scale 1:1:

> "'And then came the grandest idea of all! We actually made a map of the country, on the scale of a mile to the mile!' 'Have you used it much?' I enquired. 'It has never been spread out, yet,' said Mein Herr: 'the farmers objected: they said it would cover the whole country, and shut out the sunlight! So we now use the country itself, as its own map, and I assure you it does nearly as well" (*Carrol 1894 [2015], 168*).

The land does nearly as well, and one could easily imagine a similar absurd story of modelers aiming for an ever more detailed representation of the various systems in IAMs to realize that running the models would consume all the energy produced.

Maps abstract and idealize features of reality. Similarly, models abstract and idealize the systems they represent (cf. *Oreskes 1998*). Onora O'Neill defines the two terms as follows. Abstractions are the "bracketing" of true assumptions in representing a target system, while idealization is the invention of false things for the sake of theoretical understanding (*O. O'Neill 1996, 40*). IAMs *abstract* in leaving out aspects of the world that would be informative (e.g., a fine-grained representation of the public transport system). IAMs *idealize* reality by inventing things that are not there (e.g., a representative household). These abstractions and idealizations are done in light of the purpose of IAMs, which is to provide orientation to policymakers on a high level of abstraction. The next chapter will discuss ways to "evaluate" knowledge from IAMs regarding a specific understanding of this purpose: to provide "feasibility assessments."

One central part of IAMs providing orientation knowledge is their use of *scenarios*. Scenario analysis has been used for a long time to provide relevant knowledge to decision-makers, for instance, in military background (cf. *Wack 1985*). By characterizing the implications of different options, scenario analysis promises to condense the relevant information concerning the situation the decision-maker finds herself in. An important distinction in this relation is between *predictions* and *projections*. Predictive models have the purpose of simply estimating future developments. A weather model, for example, gives us an estimate of what the weather will be in the upcoming days or weeks. Projections of the future are different, as they rely on scenarios. Projections ask "if-then"-questions. They produce knowledge about the future based on a specific scenario and produce insights conditional to the assumptions made in the scenario. For instance, IAMs might investigate what implications it would have if we delay mitigation efforts for some time (*Luderer et al. 2013*), what impact it would have if we also pursue decisive demand-side measures in pursuing low-temperature goals (*Luderer et al. 2013*), or how the availability of different technologies like bioenergy affects climate mitigation (*Daioglou et al. 2020*). IAMs are used to ask "what-if?"-questions systematically.

We need scenarios because we need to make decisions. As the target system of IAMs is closely entangled with our actions, predicting future developments would neglect that we are free to make decisions on many aspects of reality. In this vein, one can distinguish three kinds of uncertainty that arise when modeling the climate-socio-economic future (cf. *van der Sluijs et al. 2008*; *Hawkins and Sutton 2009*). *Epistemic uncertainty* describes that we are uncertain about how the target system functions. For instance, there is considerable uncertainty about how much warming results from a certain amount of emissions or how fast certain technologies can be improved. Beyond unknowns about the system, there is further *parametric uncertainty*, which describes that our model does not match the target system perfectly. Even if we know the target system well, our simplifying models will deviate from it in various ways. Finally, there is *societal or scenario uncertainty*, which arises due to human agency and values. Since human action influences what will happen in the future, there is uncertainty relating to IAM pathways due to whether the assumed values or decisions are representative of what society, in fact, values and decides.

While the boundaries are not sharp, this last kind of uncertainty makes IAMs truly special. IAMs target societal uncertainty by designing scenarios representing a social choice concerning different courses of action. Scenarios aim to enlighten on relevant and influential decisions we have to make. For instance, if IAMs ask how the availability of new technologies can impact climate mitigation, there is a social choice in the background on whether we choose to develop and invest in these technologies. Designing a set of informative scenarios is a crucial step in such an investigation. A *scenario* is a "plausible description of how the future may develop based on a coherent and internally consistent set of assumptions about key driving forces (e.g., rate of technological change, prices) and relationships" (*van Diemen et al. 2022, 1813*). Scenarios bundle uncertainty by making broadly coherent assumptions on crucial parameters.[18] Designing a scenario also involves narrative elements, which helps to achieve coherency and relatability of the model results.

Two general classes of scenarios need to be distinguished. The first subset of projections is so-called "no policy scenarios," sometimes also called "business-as-usual" (BAU) scenarios. These scenarios describe what would happen if no additional effort to constrain emissions is made (*Clarke et al. 2014, 418*). Often, these scenarios feature as a *reference scenario* that provides a baseline against which other pathways are compared and evaluated. Overall mitigation costs are, for instance, conceptualized as the loss of GDP in a scenario compared to a baseline scenario. Such reference scenarios are useful to show where the world is heading based on already implemented policies or based on the policy plans that governments have announced. However, emission trajectories that were too high (RCP8.5) were often used as a reference, presenting overly pessimistic futures as the most likely outcome without policy intervention (cf. *Hausfather and Peters 2020*). Lower scenarios might often be a more plausible baseline (cf. *Grant et al. 2020*). Nevertheless, reference scenarios are valuable as they are the basis for calculating the "implementation gap" and "ambition gap" of current policy (cf. *UNEP 2022*; *van de Ven et al. 2023*). They play a critical role in holding governments accountable.

The second subset of projections is *mitigation scenarios*, which project different possible choices on critical decisions on climate strategy into the future. Mitigation scenarios project what "needs" to happen to reach specific goals based on different assumptions. The central scenario parameter of mitigation scenarios is the climate goal that the scenario solves, for example, the 1.5 °C or 2 °C goal. IAM scenarios translate climate goals (as well as potentially other social

---

[18] One crucial set of scenarios is the *Shared Socio-Economic Pathways* (SSPs), a set of five overarching narratives that fix certain assumptions concerning key divers such as population and education levels (*B. C. O'Neill et al. 2014*; *Riahi et al. 2017*). Each describes a way the world could develop over the 21st century. These five scenarios are the SSP1 "Sustainability (Taking the Green Road)," SSP2: "Middle of the Road," SSP3: "Regional Rivalry (A Rocky Road)," SSP4: "Inequality (A Road divided)," and SSP5: "Fossil-fueled Development (Taking the Highway)." The short narrative is translated into quantitative estimates on critical factors for climate mitigation and adaptation challenges, such as the learning curves of technologies and how technologies spread across the globe. The SSP provides assumptions on key drivers of climate change influenced by global policymaking: population growth, education levels, and others. More specific scenarios must be designed for actual IAM runs, but the SSP ensures basic comparability and connectivity for other fields.

and environmental goals) into concrete mitigation measures and strategies. This method is also referred to as "backcasting" from a desirable goal, and it is one way IAMs can be used, which I will refer to as "assessing feasibility." As the IPCC writes: "Integrated Assessment Models (IAMs) are critical for understanding the implications of long-term climate objectives for the required near-term transition" (*Riahi et al. 2022*).[19]

IAM's reliance on scenarios implies one final distinction that is important for the discussions to come. Scenarios target specific questions and make explicit assumptions about future development. I will call these explicit characteristics of the scenario *"scenario assumptions"*. For instance, if IAMs investigate the value of negative emissions, they make explicit scenario assumptions on what technologies are available, what kind of potential these technologies have, and what costs and timing one can assume. Imagine two scenarios concerning how bioenergy is available. The first might describe a future in which bioenergy runs efficiently and can be used to produce energy and various products. The other assumes that bioenergy use must be constrained (for instance, for ecological reasons) to a particular amount of land and water. One can model the implications of such constraints by comparing the two scenarios. Scenario assumptions are described and discussed explicitly in the research papers.

However, to run a particular scenario in a model, one needs to set and adjust many more parameters, routines, and input data. I will refer to this broader class of assumptions as *background assumptions*. IAMs come with a broad range of parametrizations and input data, which can be adjusted for individual model runs but are often also simply part of the overall model behavior and setting. Representing bioenergy, for instance, depends on making assumptions on the available land, the expected yield rates of crops, the technological assumptions concerning CO_2 capture rates, the energy production rates, and various institutional aspects concerning the feasibility of a socially acceptable governance scheme for bioenergy, and so on. Furthermore, a wide range of additional assumptions must be made for other aspects of climate mitigation. Such assumptions remain, for the most part, implicit in an IAM run. Modelers use and adjust the models over long periods, and trust in them builds through reliance on them in various projects.

---

[19] IAMs today fulfill many purposes on different levels of informing policymakers. For instance, Wilson et al. (*2021*) write that "Process-based IAMs are also used more directly in climate policy formulation, including the periodic global stocktake of progress under the Paris Agreement (Grassi et al. 2018), international negotiations under the UNFCCC (UNEP 2015; UNFCCC 2015), and national strategies, targets, and regulatory appraisals (BEIS 2018; Weitzel et al. 2019)" (*Wilson et al. 2021, 2*).

The exact boundary between scenario and background assumptions can be blurry at times. As modeling results depend on both, drawing the line can be difficult. In some sense, modeling results are conditional on all assumptions and parameters in the models. Giving up on the distinction, however, would make the modeling results useless, as there would be no way to understand what the models tell us in relation to the choices we have. For this book, I will thus rely on this distinction. The next chapter will discuss how to evaluate models concerning their background assumptions, whatever they are in practice.

To summarize this section, IAMs are built to inform decision-makers on the implications of various decisions on climate mitigation. IAMs abstract and idealize reality for this purpose. IAMs project various futures, which means to provide scenarios for various what-if-questions. The following section will describe how mitigation scenarios from IAMs are increasingly used to "assess feasibility."

4.6 Towards Assessing Feasibility

This section tells the rest of the story on IAMs. It describes how IAMs became tools for assessing the feasibility of climate goals and shows how impactful and central IAMs are in scientific advice on climate policy.

In 2015, international climate politics no longer lacked ambition. In December of that year, the world leaders met in Paris for the COP21 to discuss how to turn around the disappointing UN process, which has yet to see a major success since the signing of the Kyoto Protocol in 1992. The strategy was to change the strategy. Policymakers no longer tried to set emission reductions from top to bottom, fixing a carbon budget on a global level and distributing it in a shared, political effort. The new approach was collectively committing to a temperature target and letting each nation determine its ambition. Ideally, a dynamic process of increasing efforts that puts flesh of concrete mitigation efforts to the bone of the global temperature targets would unfold. Liberated from consistency between goals and actual policy, policymakers agreed on a final document, and Paris became a big success.[20] The Paris Agreement became the first internationally binding climate change agreement signed by all 192 countries. It declares the ambition to hold "the increase in the global average temperature to well below 2 °C above pre-industrial levels and to pursue efforts to limit the temperature increase to 1.5 °C

---

[20] Central to this success was the focus on a temperature goal, which did not involve direct political consequences for any nation. Early conferences focused on concrete emissions cuts, on which no agreement could be reached (cf. *McLaren and Markusson 2020*).

above pre-industrial levels, recognizing that this would significantly reduce the risks and impacts of climate change" (*United Nations General Assembly 2015*, Article 2). Paris was celebrated as "an important milestone" (*Jabbour and Flachsland 2017*) and the year 2015 was called "a visionary year when progress on global sustainability really began" (*Le Quéré and Minns 2016*).[21]

Commentators at the time, however, even disagreed concerning the achievability of limiting global warming to 2 °C, let alone 1.5 °C (cf. *Victor and Kennel 2014*; *Schellnhuber, Rahmstorf, and Winkelmann 2016*). When policymakers in Paris committed to these ambitious targets, discussions on the feasibility of these goals were everywhere. Such ambitious climate goals were not seen as viable before, as only very few modeled pathways from that time were compatible with such low levels of long-term warming. Moreover, in 2015, the world already had warmed around 0.9 °C above pre-industrial temperatures, and 2015 was the hottest year on record, with 2016 on the path to breaking this record again. Global annual emissions were still on the rise. Global warming of 1.5 °C seemed very close indeed.

Policymakers, knowing about the tension they created, included a direct request to the IPCC to "provide a special report in 2018 on the impacts of global warming of 1.5 °C above pre-industrial levels and related global greenhouse gas emission pathways" (*United Nations General Assembly 2015*). The question on everybody's mind was if 1.5 °C is feasible, which was to become one central focus of the report. Picking up the atmosphere of the Paris talks, Elena Manaenkova of the World Meteorological Organization reports: "I was there. I know the reason why it was done...[P]arties were keen to do even better, to go faster, to go even further...The word 'feasibility' is not in the Paris Agreement, is not in the decision. But that's really what it is [about]" (*Pidcock 16.08.2016*). Dyke et al. describe the atmosphere of the Paris talks as follows: "But dig a little deeper and you could find another emotion lurking within delegates on December 13. Doubt. We struggle to name any climate scientist who at that time thought the Paris Agreement was feasible" (*Dyke, Watson, and Knorr 2021*).

The request to the IPCC to produce a report on the feasibility of 1.5 °C was not seen without skepticism (cf. *Hulme 2016*). Ultimately, though, it aligned with the direction the IPCC had taken. With the physics of climate change being well understood but political action still dragging behind, Global Environmental Assessments started to focus on providing knowledge on solutions and assessing

---

[21] In September 2015, the UN General Assembly agreed upon the Sustainable Development Goals in addition to the Paris Agreement.

policy options more directly.[22] The IPCC was no exception here. Hoesung Lee, chair of the IPCC, called the primary goal of his tenure to provide "a more in-depth, and clear, understanding of the solutions" (*Lee 2015, 1006*). The request for the Special Report and the lingering feasibility questions thus fell on fertile ground. By mandate of the UNFCCC, the new report should summarize the available knowledge on warming of 1.5 °C and what the world would need to do for it to stop there.

The feasibility question was central, and Lee made it clear that the IPCC was ready to take it on: "One notion that runs through all this, is feasibility. How feasible is it to limit warming to 1.5 C? How feasible is it to develop the technologies that will get us there? ... We must analyze policy measures in terms of feasibility," Lee is quoted by Pidcock (*16.08.2016*). Answering this question and scientifically assessing policy options demands "develop and cultivate a widely accepted set of methods and tools to do so in a way that informs evidence-based policy-making" (*Jabbour and Flachsland 2017, 200*). IAMs were in place to fill a role in such an evidence-based toolbox.

Early publications on IAMs already connected pathways to questions of feasibility. In an essay on the development of IAMs in 1996, Rotmans and van Asselt, for example, describe the aim of IAMs to evaluate projections in terms of "plausibility and feasibility" (*Rotmans and van Asselt 1996, 327*). In the 2000 s, the first IAM studies directly posed the question of feasibility (cf. *Luderer et al. 2009*). PB-IAMs started to produce "mitigation pathways," scenarios that backcast the necessary steps from a given long-term temperature goal. The AR4 included these scenarios under "'safe landing' or 'tolerable window' scenarios" (*Fisher et al. 2007, 175*). The report describes these scenarios as "feasible emission trajectories and emission driver combinations leading to these [climate] targets" (*Fisher et al. 2007, 175*). At the time, however, the "assessment" of mitigation scenarios was restricted mainly to geophysical and economic aspects of feasibility.[23] IAM studies explicitly did not "deal with all kinds of societal barriers that exist in

---

[22] Jabbour and Flachsland (*2017*) analyze the 40-year history of Global Environmental Assessments (GEAs). They point out that one of the fundamental characteristics of current GEAs is a high emphasis on solutions and policy analysis. The summary and key messages of GEAs involve more solution-oriented terms such as "policy response" or "political action." Further, surveys show that decision-makers and stakeholders increasingly demand scientific input on response options and "explicit assessment of policies" (*Jabbour and Flachsland 2017, 200*).

[23] One central result of the section in the IPCC AR4, for example, is the increase of GDP losses with more stringent targets, ranging from 1 % to about 3 % by 2030 (*Fisher et al. 2007, 204*), and an estimation of the carbon prices necessary under idealized assumptions. The presentation of such results stayed short of being interpreted in terms of feasibility.

[the]formulation [of] ambitious climate policies" (*van Vuuren et al. 2007, 148*). Studies used "the term feasibility in a technical- and model-related sense," which means that "a mitigation target will be infeasible if models are not able to produce scenarios that are consistent with this target" (*Knopf, Luderer, and Edenhofer 2011, 618*). Integrated assessment studies have not yet aimed to bridge the gap to overall feasibility claims systematically. However, such attempts were already mentioned as a research question needing greater development.

Assessments moved to a more general notion of feasibility in the Fifth Assessment Report (AR5) (*IPCC 2014b*) and in related IAM studies at this time (cf. *Loftus et al. 2015; Gambhir et al. 2017; Rogelj, Popp, et al. 2018*). The AR5 introduces its chapter on "Assessing Transformation Pathways" with the question: "[W]hat will be the transformation pathway toward stabilization; that is, how do we get from here to there?" (*Clarke et al. 2014, 420*). The chapter goes on to state that a "question that is often raised about particular stabilization goals and transformation pathways to those goals is whether the goals or pathways are 'feasible'" (*Clarke et al. 2014, 420*). The method to tackle this question is by assessing "characteristics of particular transformation pathways" such as "economic implications, social acceptance of new technologies that underpin particular transformation pathways, the rapidity at which social and technological systems would need to change to follow particular pathways, political feasibility, and linkages to other national objectives" (*Clarke et al. 2014, 420*). It states: "A primary goal of this chapter is to illuminate these characteristics of transformation pathways" (*Clarke et al. 2014, 420*). The question of feasibility concerning climate goals thus found its way into the IPCC reports, and mitigation pathways produced by PB-IAMs were becoming the main tools to assess the feasibility of climate goals.[24]

However, the most direct engagement in the IPCC reports can be found in the mentioned Special Report on 1.5 °C. While only a few IAM scenarios on 1.5 °C existed before Paris, the IAM community responded quickly to the changed political landscape and produced a series of mitigation pathways in line with 1.5 °C. This scenario evidence became the basis for the feasibility chapter of the report. The central message emerged that 1.5 °C was still possible, but it would demand

---

[24] One might note that "assessing feasibility" in the AR5 was understood as a deliberative enterprise, which aimed to be sensitive to the entanglement of values and facts in the feasibility question. The chapter is influenced by the framework developed by Kowarsch (*2016*). In a paper co-authored with Ottmar Edenhofer, the lead author of working group III, they laid out the concept of "cooperative knowledge production and a role for mutual learning between experts and policymakers (*Edenhofer and Kowarsch 2015, 57*).

enormous efforts and favorable conditions and require massive amounts of negative emissions over the 21st century. The Special Report acknowledged that "no single answer to the question of whether it is feasible to limit warming to 1.5 °C and adapt to the consequences" can be given (*Allen et al. 2018, 52*). IAM pathways were seen as the tool to provide knowledge on this question.

The Special Report also shows how central IAMs have become in informing policymakers and influencing the global perspective on mitigation. The concrete implications of the feasibility assessments based on modeled pathways from IAMs were that 1.5 °C implies the need to reach Net-Zero by midcentury and that we need large amounts of negative emissions. In the aftermath of the report, many countries adopted national goals of climate neutrality around 2050 (*Climate Action Tracker 2019a*). Moreover, it was subsequently widely perceived that negative emissions are essential for reaching 1.5 °C (despite the SR1.5 already including pathways compatible with 1.5 with minimal negative emission, cf. *Grubler et al. 2018*). IAMs were central in keeping the 1.5 °C goal on the agenda by providing evidence of their attainability in the models.

IAMs gained this central role in concert with an institutional context, which they helped to shape. IAMs are versatile tools that quickly respond to policy questions and provide the knowledge demanded. Many commentators have pointed out the "performativity" of IAM scenarios (*S. Beck and Mahony 2017, 2018a, 2018b*; *S. Beck and Oomen 2021*; *Haikola, Hansson, and Fridahl 2019*). The IAM community anticipates policy questions and helps to develop a conceptual framing for the climate change problem. By providing empirical data on some solutions to the climate problem, IAM pathways make them more actionable than others, legitimizing technologies and strategies and shaping the policy discourse with their modeling results. It is thus vital to consider "how IAM results come to matter" (*Haikola, Hansson, and Fridahl 2018, 25*).

The next step was to develop more systematic frameworks for assessing feasibility based on IAMs. This was on the agenda at the time, and Rogelj, Popp, et al. (*2018*) wrote: "Policy analysts and advisors still need to translate the insights of this and other related studies into a more complete assessment of feasibility, which accounts for the broader context of societal preferences, politics and recent real-world trends" (*Rogelj, Popp, et al. 2018, 331*). A set of methods emerged that promised to assess feasibility based on empirical evidence in a systematic way, which I will get into in the next chapter. Comparing scenario data with historical analogs and systematizing expert judgments promised to "assess the relative feasibility of global decarbonization scenarios" (*Loftus et al. 2015, 108*) in a scientific way.

Summing up, this chapter introduced Integrated Assessment Models. IAMs are computer models that aim to assess human and natural systems in an integrated way. This makes them powerful and versatile tools to provide orientation knowledge on mitigation strategies and measures based on scenarios. Scenarios bundle assumptions concerning future developments. IAMs provide if-then-knowledge on the implications and interactions of certain mitigation measures and enabling or hindering conditions. IAMs are a central tool for assessing feasibility and related research questions, making up a large portion of the literature and findings of the IPCC reports of Working Group III, even though they represent only a small fraction of the whole discourse. Moreover, a few research hubs, primarily in Western states, provide most Global scenarios.

As described in Part I, feasibility is a central political concept and can be used to determine which proposals are worthy of deliberation and which can be ruled out. We often disagree on the attainability of certain desirable outcomes and the paths available for the future. This is true in the climate context as well, as the examples above suggested. Given disagreements on whether the Paris Goals are "attainable" or a "fantasy," and heated debates on what it would take to stay within their temperature guardrails, it would be clearly highly valuable to have reliable and objective feasibility knowledge in the context of climate mitigation. IAMs play a critical role in such feasibility assessment, including in the latest IPCC reports. This chapter aims to make sense of this development and engages critically with the conceptual and methodological assumptions that justify it.

This chapter starts by showing why IAMs are in a good position at all to assess feasibility (Sect. 5.1). I then argue that IAM scenarios provide scenario evidence for binary feasibility claims. This evidential relation depends on background assumptions. Existing methods of model evaluation fail to provide empirical ground to determine these background assumptions, however. I argue that, given the normative significance of feasibility, modelers need to evaluate these assumptions in light of conceptual insights provided above. These insights delimit the role appeals to the past can play and suggest to err on the side of utopian pathways (Sect. 5.2) . I then critically discuss a recent framework for assessing feasibility as a scalar concept (Sect. 5.3). Finally, I argue that feasibility assessments must be mindful of value judgments. I will introduce the perspective of values in science for the topic at hand (Sect. 5.4).

© The Author(s) 2025   87
S. Hollnaicher, *Assessing Feasibility with Value-laden Models*,
https://doi.org/10.1007/978-3-662-70714-2_5

5.1   IAMs' Special Qualities Concerning Feasibility

Chapter 4 described the shift in the IPCC and IAM community towards assessing feasibility with IAMs. This focus on feasibility emerged in light of ambitious political targets and against the backdrop of an increasing demand for science to provide knowledge on climate solutions. Contributions from IAMs are influential in this regard. By providing possible pathways for the future, they provide data on the attainability of the goals in question. For instance, the feasibility assessment concerning 1.5 °C in the IPCC Special Report was crucial in keeping the goal on the political agenda. This section explains that IAMs are in a prime spot to answer feasibility questions, as they *provide trajectories* and rely on an *overarching perspective*, both conceptual core aspects of feasibility.

Chapter 2 and 3 explicated the concept of feasibility with a particular view on the climate context. According to my extended definition, an outcome is feasible if there is a resource stock, which enables us to realize it, and which is attainable with enough certainty and without relying on unacceptable means or producing unacceptable side effects. The discussion in these chapters highlighted two core elements of feasibility: *First*, feasibility is about "viable trajectories" towards a complex social outcome in the thick sense described above. *Second*, judging an outcome to be feasible in the normatively interesting sense demands a wide-angle view that simultaneously attends to all kinds of constraints.

First, then, feasibility is ultimately about a *viable trajectory* from the status quo to the desired outcome, which can be complex and involve various diachronic steps. What is feasible is not only what we can realize with our current resources. Often, we first have to bring ourselves in the position to realize an outcome, for instance, by changing structures and institutions or developing the necessary resources and skills. When there is a path for doing so, the distant outcome is feasible even if we cannot realize it immediately. This is the first aspect: feasibility judgments on complex social goals depend on providing viable pathways.

IAMs can account for this diachronic dimension of feasibility, given that IAMs are built to provide pathways for the future and that modelers have always been concerned with their realism and plausibility (cf. *Weyant et al. 1996*; *Rotmans and van Asselt 1996, 327*). Recently, modelers have used IAMs more directly to assess feasibility. When creating mitigation scenarios, IAMs start from the outcome in question and search for paths compatible with this goal and the various constraints of the scenario. Constraints concerning available resources and processes can change over time in the models. For instance, IAMs typically include technological change, and thus, technological parameters, capacities, and cost estimates are malleable depending on other measures assumed in the scenario or model.

Models differ in representing such developments, but most IAMs involve some range of dynamic features (cf. *Krey et al. 2019*). This is essential for creating pathways that can represent the sequence of actions that are often necessary to show a challenging outcome to be feasible. IAMs generate trajectories and thus are in a prime spot to assess the feasibility of complex social goals.

Second, to determine whether these pathways are actually *viable* we must be simultaneously attentive to all the various constraints that apply to it. A goal is feasible if there is a path towards it that does not violate any constraints *when looked at them in conjunction*. At first, it might occur as if it is sufficient to check for constraints in isolation and, if we find no violation, judge the path to be viable. If such an isolated analysis suffices, it would make feasibility judgments much easier. We could consult various disciplinary experts for the feasibility of the goal or pathway regarding their dimensions. No interaction would be needed.

However, this would be insufficient, both in missing feasible paths and wrongly judging other paths as not feasible. The isolated assessments would disregard the various interdependencies between dimensions. For instance, resources from one dimension are often transformable to resources from another. Economic investments can, for instance, at times overcome a technological constraint. Complex social outcomes are only attainable when we make use of such processes. However, knowing whether these processes are available depends on whether they are not essential for other transformation processes (for instance, softening social resistance) and for other social goals (for instance, Global poverty reduction), we want to achieve. The interdependency between different goals in terms of synergies and trade-offs, what Hamlin (*2017*) calls *co-feasibility*, is essential in assessing feasibility. Co-feasibility is, on the one hand, a product of thick feasibility, as means and side-effects must be acceptable and, thus, also minimally evaluated with respect to other aims. On the other hand, it is a function of the models being relevant to policymakers and the public. The assessments must recognize interdependencies between the goals in terms of resources because policymakers are concerned with different goals. Feasibility assessment, thus, needs to attend to the set of constraints in conjunction and with an eye on other social goals. Therefore, assessing feasibility depends on having an *overarching perspective* that keeps all kinds of resources and processes in mind simultaneously.

This overarching perspective is another significant merit of IAMs. IAMs combine all the relevant systems involved in climate mitigation in an integrated fashion. IAM's key competence is to model the various interactions and co-dependencies between the systems. The economic framework of IAMs promises to translate the various implications and demands of the various systems within a common framework. Chap. 3 explained the unique role that such "all-things-considered" feasibility claims play. Moreover, IAMs started to include more of the various targets represented in the Sustainable Development Goals, thus moving away from being only concerned with climate targets (cf. *McCollum et al. 2018; Zimm, Sperling, and Busch 2018; van Soest et al. 2019*). IAMs seem to provide the overarching perspective needed for assessing feasibility, which is the second reason IAMs can assess feasibility.

In the first instance, IAMs can fulfill two fundamental presuppositions of feasibility judgments and are, therefore, in a prime spot to provide assessments on the feasibility issue. As we will see below, there are discussions on how well IAMs perform in both of these regards. To understand them, we need to explicate how model output relates to feasibility judgments about outcomes in the real world.

5.2   Model Solvability as Evidence for the Feasibility of Climate Goals

Given that the central qualities of IAMs fit well with the concept of feasibility, it is time to get clearer on *how* IAMs can contribute to assessing feasibility. This section explains the relation between solvable scenarios and feasibility as a relation of evidence.[1] This relation of evidence depends on the validity of the background assumptions. While there are methods for model evaluation, there are conceptual and methodological hurdles to relying on empirical methods to evaluate these assumptions, or so I will argue.

There is a narrow technical definition of feasibility within IAM science that we can call *"model feasibility."* A scenario is feasible "in the model," if the algorithm of the particular IAM can solve the mathematical equations representing the scenario. IAMs generally represent a scenario as an optimization problem, which minimizes overall costs subject to many constraints representing the scenario's technological, economic, and social assumptions. Scenarios are feasible in

---

[1] Conceptual contributions are often pessimistic concerning actual feasibility assessments. For instance, Lawford-Smith (*2013*) explicitly avoids discussing the feasibility of complex social outcomes, and Wiens considers feasibility assessments to be "beyond human cognitive capacity" (*Wiens 2015, 467*). As modelers, social scientists, and the IPCC, however, develop ways to assess feasibility, and given that such knowledge is precious for different agents, this book aims at a more constructive outlook on *how* climate goals can be assessed in terms of feasibility and what kind of consideration can guide our methodology.

the model sense if there is a mathematical solution to this optimization problem. Low and Schäfer describe it as follows:

> "Feasibility is a function of model solvability. [...] what is 'feasible' is de facto what is computationally possible, given initial constraints that are based on interdisciplinary and not uniformly codified expert judgments, and that change from model to model. A scenario is feasible if the model can solve for a temperature target, and a technology is feasible if it was made available as an option at all. Scenarios that are highly implausible in reality, or that produce alternative pathways to the same goal, are all technically feasible" (*Low and Schäfer 2020, 3*).

However, this technical sense of feasibility does not automatically tell us something about the feasibility of a scenario in the real world. Modelers are typically careful in pointing to the limited meaning of scenario runs. The IPCC, for instance, writes that "beyond cases where physical laws might be violated to achieve a particular scenario (for example, a 2100 carbon budget is exceeded prior to 2100 with no option for negative emissions), these integrated models cannot determine feasibility in an absolute sense" (*Clarke et al. 2014, 424*). IAM studies often caution themselves concerning far-reaching implications: "[B]ecause models are stylized, imperfect representations of the world, feasible dynamics in a model might be infeasible in the real world, while vice versa infeasibility in a model might not mean that an outcome is infeasible in reality" (*Rogelj, Popp, et al. 2018, 331*).[2]

A fruitful way to understand the relation between solvability in the model and feasibility is as a *relation of evidence*. Models produce what can be called "scenario evidence" for the feasibility of the goal and mitigation strategy in question.[3] Evidential relations are epistemic, as the presence of evidence "makes it more likely" that a hypothesis is true (*T. Kelly 2023*). When we have the right kind of and sufficient evidence for a claim, we are justified in believing it. A lack of evidence can indicate that something is, in fact, not the case if we could otherwise expect evidence to occur. On the other hand, evidence does not prove a hypothesis but only indicates its truth.

---

[2] Similar statements can be found in many IAM studies. One more instance: "These concerns need to be strictly differentiated from the feasibility of the transformation in the real world, which hinges on a number of other factors, such as political and social concerns that might render feasible model solutions unattainable in the real world" (*Riahi et al. 2015, 19*).

[3] The IPCC, for instance, writes that model-feasibility "informs," "contributes to," and "provides relevant information" (*Clarke et al. 2014, 424*) for the feasibility of different goals and trajectories for achieving these goals.

Evidential relations have certain informative features for our case. *First*, independent evidence for the same hypothesis adds up. IAM studies often seek more robust conclusions by relying on "structured scenario ensembles," which involve multiple IAMs and modeling teams investigating the same research question based on harmonized inputs.[4] Only if a portion of the models cannot find a pathway for a given scenario is taken as evidence that the scenario might be infeasible in the real world. Rogelj, Popp, et al. (*2018, 331*) "suggest that the proportion of successful scenario results can be used as an indicator of infeasibility risk."[5] The IPCC describes this as a "first, coarse indication of feasibility concern" (*Riahi et al. 2022, 379*). This practice makes sense in light of the evidential relation. If *different* models solve a particular scenario, or if other lines of reasoning suggest its feasibility, we can take this as more robust support for the feasibility of a particular goal.

Second, understanding the relation between scenarios and feasibility as evidential has another important implication. Evidential relations are not "natural" or "objective" relations in themselves, but we *take* things as evidence for a given hypothesis in light of background assumptions that we take for granted. Characterizing solvable IAM scenarios as evidence for feasibility does not preclude that it could be weak evidence or evidence that supports conclusions only subject to certain narrow conditions.[6] To know how solid the evidence is will depend on whether we can trust the models concerning feasibility. This conditionality on background assumptions gives us a tool to understand the proposed evidential relation more clearly as we need to explicate and justify these assumptions. It further connects to an observation made in the chapter on feasibility, where I argued that a conceptual feature of feasibility claims seems to be that we judge

---

[4] These so-called Model Intercomparison Projects (MIPs) allow modeling teams to learn from each other and detect individual model behavior.

[5] This feasibility relation was also proposed by Riahi et al. (*2015*): "we interpret infeasibility across a large number of models as an indication of increased risk that the transformation may not be attainable due to technical or economic concerns" (*Riahi et al. 2015, 19*). To provide an example of such a feasibility judgment relying on this sense, Rogelj, Popp, et al. (*2018, 329*) find that in "a world that promotes both geographical and social inequalities, only one out of three models attempting a 1.9 W/m^2 scenario was successful" and thereby indicating that under these circumstances 1.5 °C might not be attainable.

[6] For instance, Helen Longino writes that we take something as evidence "in light of regularities discovered, believed, or assumed to hold" (*Longino 1979, 37*). Evidential relations always depend on background beliefs and theories concerning these relations, even in the natural sciences.

things to be feasible "in view of" aspects of the context we believe to hold (cf. *Kratzer 1977*).[7]

The background assumptions concerning the evidential relation play out in the parametrization and the implicit assumptions that determine the general model behavior of a particular IAM. For instance, regarding technological change, such beliefs concern how fast specific technological parameters can change, how costs decline in build-up processes, or how much efficiency gains are achievable. For the most part, such parametrization and adjustments in the models go back to expert judgments, as the quote by Low and Schäfer above describes. They write that these adjustments are de facto "based on interdisciplinary and not uniformly codified expert judgments" and differ "from model to model" (*Low and Schäfer 2020, 3*). Such expert judgments get ingrained in the models in the continuous reliance on these models over long periods and various projects.

However, IAMs have been criticized from different angles about such assumptions. Critics point out that IAMs are overly focused on the economic and technological dimension of feasibility and are biased towards growth and technological solutions over other means of social change. IAMs are said to shed light only on "a subset of what might be possible" (*Pielke 2018, 33*; cf. *K. Anderson 2019*; *Robertson 2021*). For instance, IAMs have been criticized for excluding radical social and structural change. Moreover, build-up rates of renewable energies have outpaced even the most ambitious IAM scenarios in the past (*Creutzig et al. 2017*). Others, though, have criticized IAM scenarios as utopian and relying on overly fantastical assumptions of how fast certain techniques can be built up, for example, regarding the reliance on Carbon Dioxide Removal (cf. *Tollefson 2015*; *K. Anderson and Peters 2016*).

As integrated modeling has sought more systematic ways to assess feasibility, methods of *model evaluation* promise to investigate the validity of these background assumptions (cf. *Wilson et al. 2021*). Scholars sometimes calls such evaluations methods "model validation" or "verification" (*Trutnevyte et al. 2019*). It should be noted, however, that such terminology suggests that modelers have a specific sense of "scientific objectivity" in mind, which many reject in relation to scientific models. The meaning of "objectivity," as I will talk more about this concept below, underlying such attempts of "validating" models is that "objective" assessment "get at the objects" of inquiry. This is a sense familiar from experiments and other scientific methods. Something is objective in this sense if

---

[7] This was my second commitment for feasibility. Feasibility is relative to a particular context and involves implicit assumptions concerning this context, given which the speaker intends the judgments of feasibility to hold.

it gets a "grasp of the real objects in the world" (*Douglas 2004, 456*). However, most philosophers of model agree that we should understand model evaluation with respect to the *use* and *purpose* of the model. If this is so, we must consider whether these methods can live up to their promise in using IAMs to assess feasibility.[8]

Wilson et al. (*2021*) describes six model evaluation methods concerning PB-IAMs: historical simulations, near-term observations, stylized facts, model hierarchies, model inter-comparison projects, and sensitivity analysis. *Historical simulations* describe the ex-post modeling of a recent period. If a model can perform well in "hindcasting" the historical developments that unfolded, the background assumptions can be assumed to represent the target system well (*Wilson et al. 2013; van Sluisveld et al. 2015; Trutnevyte 2016; Fujimori et al. 2016*). A related method is to compare the results of past IAM studies to the actual development that has unfolded since the study was released. Wilson et al. (*2021*) calls this *"near-term observations"* as they compare IAM output only with the few years of data since the study's release. The third method compares IAMs to more detailed sectoral models or projections from sectoral experts. It is mainly these three methods that promise a reality check for IAMs

Results are mixed. Near-term observations have shown that while the total emission trends are within the corridor of past IAM studies (*Pedersen et al. 2021*), more detailed results reveal certain model biases. Even the most ambitious IAM pathways in the past were far below the actual build-up rates of renewable energy (*Wilson et al. 2013; Creutzig et al. 2017*), and IAM projections underestimated the role of demand decline compared to actual developments for some regions (*Le Quéré et al. 2019*). van Sluisveld et al. (*2018*) further finds that IAMs display a more substantial reliance on CCS and nuclear compared to projections by sector experts, which rely more on renewable energies to meet climate targets. While the interpretation of such "failures" is up for debate, they have been taken to argue that IAMs favor growth-based and fossil-dependent transition paths (*Robertson 2022*).

Wilson et al. caution that all methods are "limited in their ability to give confidence in IAMs' representation of modelled systems" (*Wilson et al. 2021, 8*). We often lack good data and a clear distinction between which causalities of the target system are worth reproducing and which are not. Further, as IAMs target long-term developments, near-term deviations do not necessarily indicate inadequate representation. The further methods described by Wilson et al. (*2021*) are *model intercomparison projects* and *sensitivity analysis*. Both are valuable

---

[8] It is important to note that "assessing feasibility" is *one* way how to use IAMs.

for understanding individual model behavior and making certain results more robust. Regarding background assumptions, they can be used to detect deviations from conventions within the field and understand their respective influence. As "differences between models are not systematic and models share approaches or components" (*Wilson et al. 2021, 11*), they provide only limited access to whether the models are adequate concerning our feasible limits.(cf. *Thompson and Smith 2019*).

What to make of the insights specific model evaluations provide is a contested issue. As a partly empirical question, answering it is beyond the scope of this book. There are two more general philosophical points that I want to raise. The first is that modelers must evaluate background assumptions in alignment with the conceptual content of feasibility and its normative role, at least insofar as the models aim to provide feasibility assessments of climate goals. However, often, inquiries into these assumptions use related but different concepts. For instance, studies investigate how *realistic* the models are in the sense of whether such developments are likely to occur in reality or whether they are "unprecedented." Conceptually, however, this is missing the target. Estlund's example makes this distinction between expected and feasible behavior very clear: it is extremely unlikely that he will ever perform a chicken dance in front of his class (and it is likely unprecedented), he writes, but of course, it would be totally feasible for him to do so (*D. Estlund 2020*).[9] When model evaluation or framework of feasibility assessments rely on benchmarking model output to the past, they pretend such inferences can be made.

Such "appeals to the past" risk slipping in background assumptions that are conceptually off-target. Appeals to the past are often used in a kind of "reality check" of IAM scenarios (cf. *Wilson et al. 2013, 2021*; *van Sluisveld et al. 2015*). For instance, Vinichenko, Cherp, and Jewell (*2021, 1482*) estimate the "dynamic feasibility frontier" by comparing scenarios to past episodes of fossil fuel decline. Riahi et al. (*2015*) compare IAM results to historic national emission reduction rates to check their degree of "realism." Wilson et al. (*2013*) compare technological scale-up rates in IAMs to the past.[10] *Appeals to the past* happen in the reliance on past system behavior to evaluate models, for instance, by comparing scenario

---

[9] See Brennan and Southwood (*2007, 9*) for a similar point.

[10] This method is also called "forecasting-by-analogy" (*Höök et al. 2012, 34*). Loftus et al. (*2015*), for instance, argued that historic benchmarks are "useful comparators to assess the relative feasibility of global decarbonization scenarios" and argue that such analysis should "both guide the scenario building community in constructing and testing actionable decarbonization strategies and help policy makers interpret the results of such studies" (*Loftus et al. 2015, 108*).

runs in historical simulations or near-term observations. Such methods promise an empirical ground on which to evaluate the models. I discussed the appeal of such seeming empirical estimations in describing what I called *Descriptive Feasibility*. If we consider feasibility a purely descriptive concept, scientific assessments of it must seemingly find empirical methods. Appeals to the past promise to fill this hole.

However, relying on the past is rather ill-suited to tell us something about our feasible limits. Outcomes that did not occur in the past cannot be taken as a sign of the infeasibility of these outcomes in the past. Often, something did not occur because we did not try seriously. Moreover, it is often questionable to infer that something is not feasible *now* if it was infeasible in the past.[11] Arguably, more often than not, our actions stay behind what we could have achieved, given our best efforts. This clause of conditionality to our best efforts is integral to the concept of feasibility. Feasible limits are "the limits of what people can be motivated to do" (*Wiens 2015, 453*). It seems highly questionable to me that what we have achieved in the past concerning climate change was at our feasible limits at the time. Arguably, past actions, even on a regional level, stayed behind what we could have achieved if we were as motivated as we could have been. If so, studies aiming at model evaluation by comparing IAM scenarios to historical data on emissions reductions or fossil decline rates are indeterminate regarding feasibility (cf. *Riahi et al. 2015*; *Vinichenko, Cherp, and Jewell 2021*; *Brutschin et al. 2021*).

However, worse, such appeals have a serious problem, as they involve the *moral risk* of unduly excusing agents due to their past inaction.[12] We should be especially wary not to set our feasible limits based on the past if we have good reasons to believe that the agent we addressed did not live up to her duties in the past. If we rely on the past, we might build the past unwillingness of

---

[11] It is even a mistake to conclude from outcomes that happened that these outcomes were feasible at the time, as Southwood and Wiens (*2016*) argue. Sometimes, we achieve specific outcomes only due to fortunate circumstances, something that cannot account for their feasibility. Moreover, what was feasible at some time must not be feasible now.

[12] Compare the discussion on Posner and Weisbach (*2010*). Posner and Weisbach (*2010*) argue, without empirical discussions, that only international climate treaties, which make *all* nations better off, are feasible (*Posner and Weisbach 2010*), and therefore, we should aim for a treaty that also makes high-emitting countries better off. However, this cannot be seen as a feasibility constraint for high-emitting countries since they clearly could sign a treaty outside their interest. Here, unwillingness is brought forward as a feasibility constraint, but this "is not a given constraint but rather a chosen constraint," as Roser (*2015, 81*) writes, and thus needs to be differentiated from genuine feasibility constraints (cf. *Clare. Heyward 2012*; *Caney 2014b*; *Roser 2015*; *Budolfson 2021*).

decisive agents into feasibility assessments for the future. Recall that feasibility can be used to rule out specific options. This would imply that we can then rule out (otherwise desirable) options for the future due to the inaction of the past. We end up excusing agents of ambitious moral duties simply due to their past unwillingness. This moral risk must be avoided.

While seemingly providing empirical ground for evaluating the models, appeals to the past are conceptually and methodologically ill-suited to evaluate background assumptions for assessing feasibility. The second philosophical consideration I want to raise is that there is an asymmetry between false negatives and false positives, which arises from the normative role of feasibility.[13] Again, the normative role of feasibility claims is that they are gate-keepers for our practical deliberation: only if a goal or path is feasible is it an option and thus worthy of deliberation. I introduced this asymmetry already in Sect. 3.5.

If this is true, judging a goal to be infeasible is highly consequential, as it excludes this option from further deliberation. If we are wrong in excluding an option, we have closed the door to find out more about what it involves, and if it involves costs, we are willing to accept (in light of the goal's desirability). Imagine that the scientist would have declared the 1.5 °C goal infeasible in the IPCC Special Report. This would have dismissed the political efforts and implied the scientific judgments that deliberation on this ambitious goal was not worth pursuing, a setback that could have had severe consequences.[14] Falsely judging an outcome to be infeasible is, thus, highly consequential.

The opposite error, falsely judging an outcome as feasible, does not have such high costs. It is important to note that feasibility has no normative implications beyond this gate-keeping role as I reconstructed the concept. Significantly, the fact that something is feasible gives us no reason to pursue it (cf. discussions in Sect. 3.5). Declaring a goal to be feasible only licenses deliberation on it, nothing further. Hence, there seem to be little further direct costs of including an option in deliberation beyond a potential waste of deliberative capacity.

This implies that unlike in most of science, there are good reasons to prefer false positives over false negatives concerning feasibility. A false positive is to declare a pathway feasible, but this turns out to be wrong. The consequences

---

[13] The fact that infeasible scenarios typically do not show up has been taken as a reason that the existing scenarios evidence might be biased towards more optimistic models (*Tavoni and Tol 2010*) and thus might underestimate feasibility concerns if they are not accounted for (cf. *Barker and Crawford-Brown 2013*).

[14] Arguably, the silence of the modeling community on the 1.5 °C goal previous to 2015 (cf. Chap. 4) is highly questionable in this regard, as hindered deliberation on low-temperature goals for a long time. More on this in Sect. 6.2.

would be a waste of deliberative resources, but as the decision to pursue the goal is undetermined, no direct practical consequences lurk. False negatives, however, keep options out of deliberation altogether and could result in otherwise desirable options not being discussed. Therefore, modelers are well advised to welcome a wide range of scenarios despite an increasing risk of them being infeasible in the real world. Modelers should err on the side of utopianism.

This section argued that we should understand solvable scenarios from IAMs as evidence for the feasibility of the climate goal. However, the strength of the evidence IAMs provide depends on the adequacy of the background assumptions. It is these assumptions that need to be explicated and discussed. This section further raised two philosophical considerations regarding these background assumptions: If we want to think of IAMs to provide feasibility judgments, model evaluation must avoid appeals to the past, and modelers should err on the side of including more ambitious targets, even if this increases the risk of modeling infeasible goals and pathways.

The last section will argue that paying attention to value judgments in such background assumptions is crucial. However, before doing so, I discuss a framework for assessing feasibility as a scalar, ex-post evaluation of IAM pathways that has received much attention recently.

5.3   Operationalizing Feasibility in Brutschin et al. (2021)

As modelers increased the representation of more ambitious targets despite a shrinking carbon budget, questions grew about how feasible these pathways actually are. In response, another approach to assessing feasibility has emerged. It evaluates scenarios ex-post by benchmarking scenarios "to the current knowledge regarding different types of constraints that might affect the feasibility of climate scenarios" (*Brutschin et al. 2021, 2*). This section will discuss this framework.

Assessing feasibility in a scalar sense developed over a series of IAM publications (cf. *Loftus et al. 2015*; *Gambhir et al. 2017*; *Rogelj, Popp, et al. 2018*). It found its most concrete framework in Brutschin et al. (*2021*). The AR6 devotes a whole section to this particular framework, highlighting the "important advancement" in assessing the feasibility of mitigation scenarios it provides (*Riahi et al. 2022, 381*). The authors explicitly aim to bridge "insights from the literature on the concept of feasibility [...], the IAM scenario comparison literature [...], and empirical

work pertaining to different feasibility dimensions" (*Brutschin et al. 2021, 2*).[15]
Brutschin et al. define feasibility "as the degree to which scenarios lie within
the boundaries of societal capacities for change in a given period" (*Brutschin
et al. 2021, 2*). They "stress the importance of a conceptual and operational distinction between feasibility and desirability" and highlight their empirical and
evidence-based method (*Brutschin et al. 2021, 2*).

The authors first select "a set of relevant indicators measuring decadal
changes" (*Brutschin et al. 2021, 3*) for each feasibility dimension. These indicators represent major feasibility concerns in these dimensions and are selected
so that most IAMs can provide data on them. For instance, overall costs and
carbon prices indicate economic feasibility. The decline in energy demand and
the decline of livestock share in food demand represent socio-cultural feasibility
(as they are a proxy for necessary lifestyle changes), and so on, for each dimension. In the second step, the authors propose constraints for each indicator. For
instance, the authors propose that mitigation costs beyond 5 % of GDP indicate a
"medium feasibility concern," and 10 % a high concern. In the social dimension,
a decline in livestock share in food demand is considered a medium feasibility
concern if it goes beyond 0.5 pp and a high feasibility concern if higher than
1 pp (*Brutschin et al. 2021*, Table 1). These benchmarks represent soft feasibility constraints based on "expert judgments" and the "scientific literature," as the
authors write.

When scenario data crosses these benchmarks, they are judged to involve
"feasibility concerns" in the respective dimension, ranging from low to high, represented by the numbers 1 to 3. This numerical representation allows aggregation
within and between the indicators of a dimension and visualization of their projected development over time.[16] The average feasibility concern of the indicators

---

[15] The authors refer to the scalar sense of feasibility proposed by Gilabert and Lawford-Smith
(*2012*) in the Conditional Probability Account (CPA) discussed above. I rejected this sense of
feasibility as a guide for scientific assessments, and this section will provide more concrete
considerations in this regard. However, one should note that the CPA takes scalar feasibility
to consist of the probability of bringing an outcome conditional upon a concrete agent trying. The operationalization deviates from this sense in important ways, as it is not concerned
with agents, and it remains unclear how it translates into a probabilistic sense. Brutschin
et al., though, incorporate three features of the CPA: (1) a focus on scalar feasibility, (2) con-
densation of feasibility into a single, numerical value, and (3) the value-free conception of
feasibility.

[16] In a similar fashion, Gambhir et al. (*2017*) break feasibility into three sections of low
to high "level of challenge." Rogelj, Popp, et al. (*2018*) visualize similar key metrics in a
bar from low to high "mitigation challenges." However, both studies stay short of giving a
numerical interpretation of feasibility.

within a dimension gives its dimensional feasibility concern. The average across all dimensions gives the overall feasibility concern of a scenario. This way, feasibility is made measurable based on expert judgments and the broader empirical literature on these concerns. Based on this framework, the IPCC, for instance, concludes that feasibility concerns for mitigation pathways arise mainly before mid-century, that "institutional feasibility challenges appear to be the most relevant" (*Riahi et al. 2022, 382*), and that the "reality check shows that many 1.5 °C compatible scenarios violate the feasibility corridors" (*Riahi et al. 2022, 382*).

As an *"operationalization"* of feasibility, the framework represents a scientific tool that provides a temporary definition of a concept with the goal of making it measurable (cf. *Feest 2005*). Feest (*2005*) writes that such methodological conceptualizations are not meant as a semantic thesis but as provisional tools to gain scientific insights into a specific problem.[17] The authors aim to increase precision and comparability in assessing the feasibility of different scenarios, for the first time evaluating feasibility "in a systematic way," as they write (*Brutschin et al. 2021, 1*). Moreover, the framework explicitly aims to cover all relevant dimensions of feasibility, including previously neglected social and institutional dimensions. The IPCC welcomes this development as an "important advancement since social and institutional aspects are as if not more important than technology ones" (*Riahi et al. 2022, 381*). Notably, the goal is to make feasibility measurable and put it on empirical ground.

However, this last goal is questionable if we look at the operationalization in more detail. *First*, it relies heavily on appeals to the past in estimating constraints for future action. Such appeals to the past are often empirically questionable. Take, for instance, the proposal in Brutschin et al. (*2021*) to compare mitigation costs with public spending in the COVID crisis. It is a wild guess if the public spending at the time can tell us much about our feasible limits. Imagine, for instance, that the pathogen would have been more dangerous and the containment measures thus more severe. In this case, would we not have seen more public spending to salvage society and the economy? If so, the spending at the time was not at the limit of what was practically possible. The whole situation allows for an interpretation that points in a different direction. What we witnessed was

---

[17] Feest writes: "in offering operational definitions, scientists were *partially and temporarily* specifying their usage of certain concepts by saying which kinds of empirical indicators they took to be *indicative of* the referents of the concepts" (*Feest 2005, 133*). Scenario feasibility, as proposed, is meant to be such a tool: the authors see it not as "a final judgement on feasibility but rather as a tool to map out areas of concern and highlight enabling factors which can mitigate them" (*Brutschin et al. 2021, 3*). The main aim of such methodological steps is to "get empirical investigations' off the ground'" (*Feest 2005, 134*).

that the *perception* of what is feasible turned out false, as such significant public spending was surely seen by many as infeasible before, a belief quickly reversed in light of an acute and imminent crisis. Thus, it remains an open question, what spending would be feasible if climate change would be taken seriously as a threat to humankind. Analogical reasoning based on the past cannot provide good empirical evidence for our feasible limits. Moreover, as argued above, such appeals are conceptually mistaken and run the risk of excusing capable agents simply due to their past unwillingness. Lowering our normative expectations on grounds of past inaction is a moral failure we should try to avoid. What Brutschin et al. (*2021*) describe as deriving constraints from "scientific literature" are such appeals to the past. This is the first point contra Brutschin et al. (*2021*).

The *second* objection is that the framework promises an empirical and scientific estimation of feasibility but, on closer look, involves unreasonable levels of uncertainty. There are reasons to believe we cannot reduce this uncertainty on empirical grounds alone. The uncertainty is mainly filled with "expert judgments," giving few experts from a particular field undue influence on normatively significant knowledge. As I will argue in the next section, this uncertainty is an entry point for implicit value judgments in such assessments.

Let me run through two examples from Brutschin et al. (*2021*) to illustrate this. The literature the authors cite concerning concrete thresholds is mostly vague or inconclusive, once we take a closer look. Take, for instance, the technological feasibility dimension, for which Brutschin et al. (*2021*) define the "medium concern threshold" for decadal increase in wind and solar share of electricity generation at 10 pp and the "high concern threshold" at 20 pp. Scenarios with a wind energy build-up beyond 10 % are thus evaluated as a medium feasibility concern, citing "Own analysis; Wilson et al (2020)."[18] However, Wilson et al. (*2020*) is a paper arguing for the advantages of more granular technologies. It is hard to guess from which part of the paper one can derive any concrete numbers concerning constraints on renewable build-up.[19] Deriving any concrete estimate from the one possible graph seems bold, especially as the mid-range unit size, presumably applicable for wind and solar, is represented by technologies such as "plastic boating," "organic pesticides," and "SO_4 turpentine." What this tells us about the feasible limits of wind and solar diffusion in times of planetary crisis

---

[18] Some authors (for example, Wilson and Grubler) appear in most studies cited on thresholds and other empirical sources being previous IAM studies. The framework makes feasibility claims dependent on expert judgments from a small part of the modeling community.

[19] Presumably, the authors rely on the "diffusion times" of different technologies. However, the respective graph (B) shows a correlation between more rapid diffusion and smaller unit size, which appears to be very weak.

is far from clear. How Brutschin et al. (*2021*) derive any specific numbers from this paper, one can only guess.

Even in cases where the literature is actually targeted at proposing feasibility constraints, it appears questionable if the literature cited in Brutschin et al. (*2021*) can provide good enough evidence for the role assigned. The geophysical dimension of feasibility is assessed by comparing wind and PV production with the "global potential." Determining such potential is, however, fraud with conceptual and scientific uncertainty. What counts as a realistic potential depends on implicit feasibility considerations, for instance, how much land is available for wind energy. In the cited study of Deng et al. (*2015*), agricultural areas are made available for wind at 3 to 20 %, depending on the scenario. Other parameters in such studies have similar extensive ranges of uncertainty. Deriving any concrete constraint from such studies will involve many value-laden feasibility judgments in itself. A particularly contested constraint concerns the indicator carbon prices, for which Brutschin et al. (*2021*) propose $60 and $120 as thresholds based on "own analysis." The paper does not go into further details. However, any determination of a "feasible" carbon price involves complex political feasibility judgments and will highly depend on how one envisages the political implementation of the actual climate policy.

I would not belabor this point and the specific examples if it were not consequential. A sensitivity analysis in a recent study showed that a 25 % lower feasibility threshold results in a reduction of overall feasibility concerns by up to 4 points on a 1 to 7 scale, thus making the difference between "no concern" and "medium concern," or "medium concern" and "high concern" (cf. *van de Ven et al. 2023*). Given the mentioned uncertainty, 25 % seems relatively small compared to the overall uncertainty. Moreover, the analysis of Brutschin et al. (*2021*) is given a whole section of the IPCC AR6 and welcomed with great promise as an expert-based, graphically appealing, and numerically comparative feasibility estimation of modeled pathways. Biases in such authoritative assessment of feasibility by the IPCC are problematic, as they might narrow the public discourse prematurely and risk smuggling in value judgments. Given the standing of integrated modeling within the IPCC, wrongly calling specific pathways or goals less feasible could foreclose debates on morally more desirable paths and goals.

Thus, we should reject accepting such a framework as an instance of empirically grounded feasibility assessments. It involves conceptually fraud appeals to the past and the moral risk of excusing us due to past inaction. It further involves high uncertainty, leaving much room for implicit, value-laden judgments. The main benefit of the framework is that it helps to explicate background assumptions in assessing feasibility with the models formerly implicit in IAMs.

Moreover, it can serve us as an evaluation method for pathways that depends on a particular value outlook (cf. my discussions on claims on scalar feasibility in 3.5).

The worries concerning the framework of Brutschin et al. (*2021*) align with recent philosophical criticism on the treatment of feasibility in the social sciences. McTernan (*2019, 36*) argues that we should be "epistemically modest" in assessing feasibility since the social sciences generally do not provide the right kind of knowledge to make such claims. Knowledge from the social sciences is not robust and general enough to be able to rule out political proposals, and it does not provide the causal knowledge necessary to assess soft constraints. Boran and Shockley (*2021*) warns that operationalizing feasibility along the lines of the CPA might be applicable in "limited scale decision-making processes" but that "it does not capture the formidable complexity of climate change" (*Boran and Shockley 2021, 36–37*). Finally, Schuppert (*2021*) warns of the risk of masking uncertainty and reproducing privileges and past injustices in feasibility assessments.

This section discussed the influential operationalization of feasibility as proposed by Brutschin et al. (*2021*), which compares scenario data to externally derived feasibility constraints. The framework is an explorative tool that may help in making implicit background assumptions explicit but it does not provide an empirical way of assessing feasibility. Conceptually, appeals to the past risk building past inaction into feasibility judgments for the future, and epistemically, the framework reveals unbearable uncertainty that would make it ill-advised to derive normatively consequential knowledge from it. The following section will argue that debates on background assumptions and feasibility constraints, in general, involve value judgments. Modelers should pay closer attention to value influences when assessing feasibility.

5.4   The Need to Focus on Value Judgments

This final section will argue that feasibility assessment must pay more attention to value judgments. This section briefly introduces the discussions of values in science as a background. It clarifies what it means to speak of value judgments. This provides the basis for the next part of the book to investigate the normative dimension of assessing feasibility with IAMs.

As the discussions above reveal, assessing feasibility on empirical grounds alone is unavailable. The concrete attempts revealed that they often could not fulfill the conceptual requirements that feasibility carries and involve fraud methodologies in this regard. One of the central problems of relying on appeals to the past and highly uncertain methods is that they are entry points for implicit value judgments. The emphasis to ground feasibility assessment in empirical methods tend to *mask* these values. This section argues that it is crucial to consider the value dimension more explicitly when assessing feasibility. The background for the emphasis that Brutschin et al. put on "empirically assessing feasibility" is arguably a more general view of science as value-free (cf. the discussions on *Descriptive Feasibility* in Sect. 3.1.)

The goal of *objective* science is often equated with being value-free. One sense of objectivity provided by Douglas (which she rejects) is that "all values (or all subjective or"biasing" influences) are banned from the reasoning process" (*Douglas 2004, 459*). Such a sense can be found in statements by modelers, for instance, when they take the meaning of the IPCC mandate to imply that "non-epistemic values [are] [...] an unacceptable element of the process of providing a knowledge assessment" (*Gundersen 2020, 100*). This is a direct expression of what is known in the philosophy of science as the "value-free ideal." Heather Douglas (*2009*), who was influential in bringing the ideal back into philosophical debates, defines the ideal that "in the heart of science, at the moment of inference, no social and ethical values were to have any role whatsoever" (*Douglas 2015, 611*).[20]

The established term for analyzing and discussing undue influences is *"value judgment."*[21] Generally, value judgments might be understood as any judgment that contains some evaluative or normative element. *Values* provide reasons for making choices in the scientific process one way or the other. Valuing something means we have reasons to pursue or enhance it (cf. *T. M. Scanlon 1998; Rowland 2019*). For instance, if we value simplicity, this gives us reasons to pursue a simpler theory over a more complicated way of explaining a phenomenon. This means that values often make a difference in the scientific work itself and thus can be detected in so-called "epistemically unforced" decisions made in the scientific process. However, value judgments are somewhat of an umbrella term in the philosophy of science. It ties the seemingly problematic influences rooted

---

[20] Cf. earlier discussions on the issue in Weber (*1904*).

[21] The term itself goes back to Thomas Kuhn's influential paper "Objectivity, value judgment, and theory choice" (*Kuhn 1977*). As guidance in science comes down to values, Kuhn argues, scientific disagreement can sometimes persist despite a shared set of norms, criteria, or values that different scientists subscribe to (and in light of the same evidence). This is nicely explainable by values, which need to be interpreted and weighted for guidance. The values Kuhn was alluding to were "epistemic values," such as empirical accuracy, consistency, scope, simplicity, or fruitfulness (*Kuhn 1977, 357*), not social, ethical, or political values.

in interests, worldviews, ideologies, religions, political positions, etc. A common element is that these "values" cannot themselves be justified empirically but answer only to value-based reasoning and thus seem to be in tension with objective science.[22]

Philosophers of science typically make two distinctions when discussing the value-free ideal of science. First, even defenders of the value-free ideal accept so-called *epistemic values* in science. Values such as empirical accuracy, consistency, or simplicity are essential to scientific practice and not in tension with the ideal of science as objective (cf. *Steel 2010; Douglas 2009, 93*). The value-free ideal concerns other kinds of values, such as ethical, social, and political values. These are known as *nonepistemic values*, or "contextual values" (*Douglas 2009*). It is their influence that seemingly undermines the objective and neutral character of science. Examples of nonepistemic values are equality, justice, sustainability, profitability, and, on the darker side, racist or sexist values.

A second distinction within the ideal is between the *"context of discovery"*, where values play legitimate roles, and the *context of justification*, which needs to be kept value-free (cf. *Lacey 1999*). Nonepistemic values are legitimate and essential for shaping the research agenda and selecting questions scientists pursue. As science is after "significant truths," as Kitcher (*2001*) puts it, such influences are legitimate. For instance, concern for safe planetary conditions can and should inspire research into the climate and ecosystems and our response strategies to the current crisis. Further, values can legitimately constrain methodologies, for instance, when ethical guidelines apply to how researchers can treat human or nonhuman subjects in experiments (*Elliott 2017*). The value-free ideal holds, however, that the inner stages of the scientific process, such as assessing data, testing hypotheses, and scientific reasoning, must be free of nonepistemic value influences.

Given these two distinctions, the value-free ideal is the demand that nonepistemic value judgments have no role to play within the context of justification. However, the ideal has seen severe criticism in recent decades, to the point that "the debate over whether non-epistemic values can play a legitimate role

---

[22] Some have recently argued that the term "value judgment" is not a good fit for some of the influences we are concerned with. Biddle (*2013*) argues that other factors are of concern as well: norms, subjective preferences, and ideological assumptions, of which some are not "properly described as values" (*Biddle 2013, 132*). Hilligardt (*2022*) argues that value judgment covers three distinct phenomena: interests, perspectives, and opinions. She argues that this risks conflating the three, which demand different response strategies (*Hilligardt 2022, 4*). However, one might hold onto the term nevertheless, as all three share a need for normative justification, which, for the moment, can be lumped together.

in science has largely come to a close," as Holman and Wilholt (*2022, 211*) write. Most philosophers of science agree that the value-free ideal cannot and should not be attained. *First*, the distinctions and defenses of value freedom have been shown to be insufficient for keeping science value-free. The distinction between epistemic and nonepistemic values is not a clear boundary. Choices concerning the epistemic values that guide scientific work can involve nonepistemic value questions (cf. *Longino 1995*). Moreover, value judgments in the context of discovery often "spill over into that of justification" (*Carrier 2021, 7*) when, for example, research agendas influence what kind of evidence is produced and thereby influence what kind of theories find support (cf. *Elliott 2017, 59*). *Second*, philosophers of science have brought forward a range of arguments that show how nonepistemic values are actually essential even in the inner stages of science. Three lines of arguments are most prominent: the argument from inductive risk (cf. *Rudner 1953*; *Douglas 2000, 2009*), the reliance on thick concepts in studying social phenomenon (cf. *Dupré 2007*; *E. Anderson 2002*; *Alexandrova 2017*; *Abend 2019*; *Djordjevic and Herfeld 2021*), and the underdetermination of theory by the available evidence (cf. *Longino 1990*; *E. Anderson 2004*; *Biddle 2013*; *M. J. Brown 2013*).

The value-free sense of objectivity is, thus, elusive and masks the widespread reliance of science on value judgments. This is arguably the case in the recent attempts of assessing feasibility on seemingly empirical methods. When modeling pathways for the future and using them as evidence for the feasibility of climate goals, a range of value judgments arise, as the next chapter will show.

5.5   Summary

This chapter first provided two conceptual presuppositions to apply IAMs to questions of feasibility. Feasibility demands viable trajectories and an overarching perspective. Both aspects are strengths of IAMs, and thus, IAMs are in a good spot for assessing feasibility. I argued that we should think of solvable models as scenario evidence for the feasibility of the climate goal in question. However, we must consider the background assumptions supporting this evidential relation. Methods of model evaluation promise to investigate these assumptions on empirical grounds, but the methodologies used miss the target. Appeals from the past cannot determine our feasible limits for the future and risk excusing agents due to their past unwillingness. High uncertainty, moreover, makes these methods prone to slipping in implicit value judgments. The promise of epistemic evaluation to provide "objective" assessments of feasibility is, thus, questionable. The sense of objectivity here seems to be alluding to value-free science. I explained the background of this ideal and how it generally came under criticism recently. We must thus investigate the models and methodologies of assessing feasibility in light of value judgments. The next part of the book will dive into the normativity of modeling pathways. It will show that various value questions arise when modeling different climate futures. Modelers need to deal with these value questions to provide legitimate and objective assessments with IAMs.

Part III
The Implicit Normativity of IAMs

Value Judgments in IAMs

The first part argued that feasibility has an important value dimension, as feasibility judgments are always made in view of background assumptions involving normative aspects. The previous chapter showed that such background assumptions are also a part of feasibility assessments and that a purely empirical understanding of them fails due to methodological and conceptual issues. Thus, we need to consider the value dimension of feasibility assessments with IAMs more fully.

This chapter provides a taxonomy of value judgments in IAMs. So far, there is little comprehensive analysis of value judgments in PB-IAMs. Existing contributions to the issue often focus either on Cost-Benefit models (cf. *Gardiner 2011*; *Schienke et al. 2011*; *Frisch 2013, 2018*; *Budolfson et al. 2017*; *Frank 2019*; *Mintz-Woo 2021b*), are restricted to a particular aspect, for instance, the discount rate (cf. *Caney 2009*; *Moellendorf 2013*), or stay rather cursory in their treatment of the value issue (cf. *Weyant 2017*). While PB-IAMs share some value assumptions with CB-IAMs, they play out differently in this context (cf. *Rubiano Rivadeneira and Carton 2022*). Discussions of the implicit ethics of PB-IAMs are, thus, urgently needed. As Tavoni and Valente (*2022*) state, "the normative components of models—more than the physical and socio-techno-economic ones—are the most fraught by uncertainty and yet the least understood. We suggest a research agenda to explore uncertainties of evaluation frameworks, transcending the current implicit normativity of IAMs" (*Tavoni and Valente 2022, 1*). This chapter is a contribution to this task.

It explicates the most crucial value judgments in assessing feasibility and integrated modeling in general, starting with explicating ethical questions that arise when choosing indicators and benchmarks for feasibility (Sect. 6.1). It then shifts

© The Author(s) 2025
S. Hollnaicher, *Assessing Feasibility with Value-laden Models*,
https://doi.org/10.1007/978-3-662-70714-2_6

the focus to the models themselves, discussing value judgments in the agenda-setting (Sect. 6.2), the general framework (Sect. 6.3), the concept of well-being (Sect. 6.4), the representation of inequality (Sect. 6.5), the choice of a discount rate (Sect. 6.6), the choice of domain (Sect. 6.7), and finally the handling of uncertainty in parameters and technological assumptions in IAMs (Sect. 6.8). This chapter explicates the value question modelers face for each aspect and provides discussions and alternatives to the existing value judgments embedded in scenario evidence from IAMs.

6.1   Choosing Feasibility Indicators and Thresholds

This section picks up the methodological discussions of the last chapter in explicating the value judgments that arise in choosing indicators and thresholds for feasibility analysis. It argues that particular indicators and constraints favor the interest of specific agents over others. As these choices cannot be determined on empirical grounds, as the last chapter argued, making them involves value judgments.

When IAM scenarios are used as evidence for the feasibility of a particular goal or pathway, modelers need to decide what kind of constraints they apply within the model or ex-post in assessing particular pathways. The last chapter argued that determining feasibility constraints on purely empirical grounds faces conceptual challenges and, when looked at in detail, involves high uncertainties. The chapter argued that relying on a putatively epistemic method will mask the fact that these decisions involve value judgments. This is the first area of value judgments.

It is helpful to focus on particular examples from the framework of Brutschin et al. (*2021*) in order to understand the value judgments, as they are here most explicit. Previous studies exploring feasibility with IAMs involves similar methods and, subsequently, equally give rise to value judgments. Brutschin et al. assess feasibility by comparing scenario data from IAMs on representative indicators to thresholds they provide based on the literature and expert judgments. For instance, the economic dimension of feasibility uses data on the following four aspects:

- *Carbon prices*, as the absolute levels of carbon taxes needed in the model to meet the target in question.
- *GDP losses*, as a representation of the overall mitigation costs, derived from a comparison with a baseline scenario.

- *Energy investments,* as the ratio of additional investments compared to a baseline scenario.
- *Stranded coal assets,* as the "share of prematurely retired coal power generations" (*Brutschin et al. 2021,* Table 1).

Data on these indicators is compared to the proposed thresholds to derive a numerical value of (economic) feasibility concern for a scenario. For instance, carbon prices of $60 are taken to indicate medium feasibility concern, and $120 high feasibility concern. GDP losses of 5% and 10% indicate medium and high feasibility concerns, respectively. Ratios of additional energy investments beyond 1.2 and 1.5 are of medium and high concern. Finally, if more than 20% of coal assets in a scenario are retired prematurely, this indicates medium concern and more than 50% a high concern. These benchmarks are used to calculate the relative economic feasibility of the scenario in question.

While each indicator has a plausible narrative for being concerned with feasibility, they also touch on substantive value questions. The amount of stranded coal assets indicates the size and speed of the infrastructure overhaul of a particular mitigation scenario. Retiring infrastructure prematurely will imply challenges and various consequences for society. For instance, coal companies could be owed compensation. A premature closing of coal plants could lead to job losses, which could create social resistance. Societies could suffer from fatigue of change if too much changes too quickly. However, it is crucial to see that these particular methodological choices also involve substantive value judgments in the sense of implicitly promoting particular interests and worldviews over others.

Consider the example of stranded coal assets as an indicator. Choosing this particular indicator is not value-neutral. *First,* it is not neutral with respect to interests, as it implicitly promotes the interests of the coal industry, coal suppliers, and coal workers over other interests. Stranded assets are a financial risk to companies that run coal plants, and thus, it is in their interest to choose mitigation strategies that involve later coal exit dates. Picking this indicator will support this, as pathways with early coal exit then appear less feasible, and this may be taken as a reason to dismiss pathways that are too challenging in this regard in our deliberation.

Second, choosing this indicator is not neutral concerning different policy options. Scenarios involving earlier coal exits appear less feasible, and this choice will implicitly advantage policy solutions which are compatible with a later end to fossil dependence. For instance, policy pathways that shift mitigation burdens to the future by using carbon removal techniques appear more feasible. Further, this indicator promotes policies such as CCS over policies that replace coal with renewables. IAMs might not be "policy-prescriptive in a strict sense, but they are certainly policy-shaping to a degree beyond policy relevance," as van Beek et al. (*2022, 200*) puts it.

Third, it makes no distinction *whose* assets are prematurely retired and thus treats coal plants in rich and potent countries similar to plants in the Global South. If climate pathways demand the premature closing of fossil plants, this could have very different value implications in different regions and highly divergent normative justifications. Different value outlooks are promoted depending on whose assets are promoted to the status of a feasibility risk. Treating all coal assets equally involves a value judgment, as would applying any other criteria.

Alternatively, take a second example in the indicators of mitigation costs, which is a central data point of IAMs. Conceptually, there is a clear distinction between a costly (even highly costly) path and an infeasible path (*Southwood 2018, 2*). Just because something involves extreme costs does not imply that the measure is not feasible. For instance, it might be highly costly for Donald Trump to pay back all his evaded taxes, but this does not make it less feasible for him to do so. Defining overall mitigation costs as an indicator for the feasibility of different mitigation pathways and proposing a threshold for it gives rise to value questions, as it makes a judgment of what amount of costs are acceptable. This is a political and ethical question, however.

This is not to argue that the chosen indicators and threshold are indefensible. In contrast, it seems highly valuable to explicate what particular scenarios involve regarding stranded assets and mitigation costs to deliberate on these pathways. However, making these methodological decisions raises complex political and ethical questions. Constraints on such indicators cannot be chosen on empirical grounds but must be discussed with value questions in mind. The authors' statement to provide an empirical assessment of feasibility that retains a "conceptual and operational distinction between feasibility and desirability" (*Brutschin et al. 2021, 2*) is misleading, as the choices made involve judgment on both.

Brutschin et al. (*2021*), thus, provides a relatively straightforward example. However, similar value judgments also appear within the models. Modelers must define some internal thresholds at which a scenario run is considered unsolvable in the models. Models need to set boundaries to scale up rates of technologies, maximum carbon prices, and similar aspects. Defining these cut-off points involves value-laden choices. The last chapter proposed to err on the side of producing more speculative scenarios instead of excluding options prematurely because they appear too challenging. Wherever one draws the line, however, will involve value judgments. Feasibility constraints are not purely empirical but involve value judgments on acceptable means and side effects (cf. Chap. 3).

Finally, modelers must decide what kind of scenario data to report. These decisions will be made in relation to what is perceived to be relevant for the deliberation on the feasibility and desirability of various futures. Choosing what kind of scenario evidence is reported involves value-laden choices, even if modelers refrain from using them as direct feasibility indicators.

To conclude, choosing indicators and thresholds involves value judgments, as I explained with the examples from Brutschin et al. (*2021*). Similar value questions appear within the models, as modelers have to define internal constraints and decide what data to report.

6.2   Agenda-Setting: Producing Evidence and Ignorance

A second set of value questions arises in the agenda-setting stage of integrated modeling. Even defenders of the Value-free Ideal acknowledge that nonepistemic value judgments must play a legitimate role when researchers decide on a topic and particular research question. However, sometimes, value judgments in the agenda-setting phase "spill over" (*Carrier 2021*) to other areas. This section will show that value-laden decisions in agenda-setting can have various policy implications in need for explication.

IAM pathways are potent visions of the future, which influence policy decisions and feature prominently in the reports by the IPCC. This prominence implies that whatever kind of IAM pathways are modeled, they have some policy implications and can make specific public options appear backed up by scientific knowledge. Certain kinds of futures become politically more "actionable" than others, simply by there being scenario evidence from IAM regarding them. This has been described as the "performativity" of mitigation pathways (*S. Beck and Mahony 2017*; *McLaren and Markusson 2020*; *van Beek et al. 2022*). IAMs are said to have "world-making power […] by providing new, political powerful visions of actionable futures," as S. Beck and Mahony (*2018a, 1*) puts it.

Philosophy of science provides the lens of *agnotology* to this issue. Agnotology is the study of ignorance and doubt in science (*Proctor and Schiebinger 2008*). Ignorance can be purposefully created, for instance, when fossil companies spread doubt about climate change to delay action (cf. *Oreskes and Conway 2010*; *Biddle and Leuschner 2015*). At other times, ignorance is simply the "unintended by-product of choices made in the research process" (*Kourany and Carrier 2020,*

4).[1] These latter cases can be applied to IAMs. It often involves certain value judgments.

One particular case of ignorance as a passive construct can be witnessed in the history of the 1.5 °C goal. When the Paris Goals were adopted in 2015, policymakers and scientists realized that there was little knowledge on whether and how this goal was achievable. Almost no emission trajectories from IAMs were compatible with 1.5 °C at the time. The AR5 (*IPCC 2014b*), published just a year earlier, lacked discussions of 1.5 °C completely, as the information was deemed "too few, and therefore there was no clear scientific basis on which to assess the 1.5 goal," Livingston and Rummukainen (*2020, 12*) writes. The AR4 in 2007 had no scenario below 2 °C, and only 6 out of 177 scenarios were in the range of 2 to 2.4 °C (*Clarke et al. 2014, 430*). A meta-analysis done in 2007 writes that "[s]tudies which investigate the costs of deep mitigation, e.g. more stringent stabilization targets such as 450 ppm CO_{2-eq} or lower, are very scarce as these targets are generally considered to be infeasible" (*Barker and Jenkins 2007, 4*).[2] How could climate economics completely miss the 1.5 °C goal just before it was put into international law by the leaders of the world?

One possible explanation, of course, is that this goal really was infeasible. Then, IAMs would have been right not to model pathways. However, this answer is unconvincing. At the time, a substantial carbon budget was still left for 1.5 °C. Behind the presumption of infeasibility lurk a range of value-laden factors that go back to agenda-setting and commitments of the research community. Climate economics focused at the time on cost-benefit analysis of climate change, which perceived much higher emission trajectories to be "economically optimal" (cf. *Nordhaus 2013, 2017*). This normative conception of economics also influenced the knowledge of feasible visions for the future as pathways produced by CB-IAMs were also used to inform what kind of futures were available. As CB-IAMs mostly produced pathways above 2 °C, it appeared as if lower trajectories were not feasible. This goes back to CB-IAMs involving a particular value perspective: a focus on optimizing economic growth in terms of consumption and a tendency

---

[1] Kourany and Carrier (*2020, 13–15*) distinguish four ways (and give examples) in which such unintended ignorance can arise: (1) ignorance resulting from a particular *definition* of the research problem, (2) ignorance resulting from the *conceptual framework* available to the scientist, (3) ignorance resulting from *choice of methodology*, and (4) ignorance resulting from a *biased composition* of the researcher community, which focusses only on particular issues.

[2] The set of *Representative Concentration Pathways* did not include pathways that would limit global warming to 1.5 °C either, when introduced in 2011 (*van Vuuren et al. 2011*). Modelers only added such a pathway later.

to rely on technological, gradual, and supply-side solutions instead of rapid transformation and demand-side interventions (cf. *Creutzig et al. 2017*; *Hare, Brecha, and Schaeffer 2018*). These value judgments led to lower temperatures appearing undesirable in the models.

Moreover, among economists at the time, scenarios on low temperatures did not pose "exciting science," as Tol (*2007*) bleakly suggests.[3] For economists, the 1.5 °C goal "was not deemed scientifically interesting" (*Livingston and Rummukainen 2020, 12*). The field instead rewarded determining globally optimal targets instead of modeling pathways on low-temperature goals. Results that challenged the conventional wisdom of climate scientists (which often advocated for low warming levels) were scientifically more interesting. Another factor also comes into play here: a biased composition of the researcher community (cf. *Kourany and Carrier 2020, 15*). The described lack of attention to lower warming targets coincides with the dominance of researchers from the Global North in this field. It was voices from the Global South that put 1.5 °C on the political agenda, but such voices play only a marginal role in IAM research. Research agendas were arguably shaped by the interests of rich and less vulnerable nations, which paid closer attention to the EU's goal of 2 °C and higher goals than the 1.5 °C goal. As the general composition of the research community has not changed considerably, such value aspects can be expected to still influence the kind of knowledge IAMs produce.

The kind of pathways that emerged confirm this suspicion. With the adoption of the 1.5 °C goal, modelers started producing many scenarios for this target, and the special report included 90 scenarios on this previously deemed infeasible scenario (*Masson-Delmotte et al. 2018*). These pathways relied heavily on technological solutions and shifted burdens to the future and, in effect, to the Global South by relying on massive amounts of bioenergy with carbon capture and storage (BECCS). Such reliance on BECCS is arguably another instance of ignorance as a passive construct, as the particular methodologies and worldviews in place led to knowledge production only on a small (value-laden) section of all

---

[3] Tol discusses Europe's adoption of the 2 °C goal and argues that policymakers lacked any scientific basis for the goal. In contrast, he highlights that the "technically sound" amongst the existing studies "argue that it is not in our collective best interest to stabilise concentrations—unless there happens to be a cheap, large-scale, carbon-free energy source—let alone at the levels needed to meet the 2 °C target" (*Tol 2007, 430*). The problematic background of Tol's arguments is, for example, discussed in Gardiner (*2011*).

visions for the future for 1.5 °C (cf. *Geden 2015*; *K. Anderson and Peters 2016*; *S. Beck and Oomen 2021*; *Hollnaicher 2022*).[4]

What kind of knowledge and ignorance exists is often influenced by nonepistemic values and interests. Such ignorance as a passive construct can be influential in shaping policy discussions. This has been the case regarding the 1.5 °C and the kind of pathways that emerged after its adoption. This implies that research priorities must be discussed with value questions in mind (cf. *Lacey 1999*; *Kitcher 2001*). IAMs are powerful tools that inform on feasible future that can limit what options we deliberate. The research agendas of IAMs should be seen as a public resource in need of critical discussions.

6.3 Cost-Effectiveness and Burden Sharing Principles

While the previous section discussed value judgments in applying IAMs to the feasibility question and in the agenda-setting phase, the sections to come will dive into specific value judgments arising within the models. PB-IAMs follow an approach called "Cost-Effectiveness-Analysis (CEA)," which means they aim to find the least-cost pathways for a given climate goal and set of scenario assumptions.[5] Mostly, IAMs rely on a Social Welfare Function (SWF), which codifies many value judgments in the models. This section argues that the Cost-Effectiveness Approach is a central value judgment in IAMs and that it can be contrasted with other principles of distributing burdens.

We can first acknowledge that CEA is a major step towards *restricting* the influence of value judgments compared to Cost-Benefit models. In CEA, all ethical questions concerning climate impacts are no longer subject to economic optimization within the models but are transferred to the scenario definition. In "CEA [cost-effectiveness analysis], value judgments are to a large extent concentrated in the choice of climate goal and related implications," the IPCC writes (*Masson-Delmotte et al. 2018, 150*). Nevertheless, cost-effectiveness

---

[4] In response to criticism, the IPCC included one such deep mitigation scenario relying on ambitious demand-side reductions with Grubler et al. (*2018*). However, the general message on the feasibility of the 1.5 °C remained largely untouched (cf. *Robertson 2022*).

[5] There are different solving methods in different PB-IAMs. However, most models involve a welfare function that it maximizes, either globally (in "welfare-optimizing models") or maximized for each world region (in "partial/general equilibrium models"). For example, the ReMIND model's core is "a Ramsey-type optimal growth model where intertemporal welfare is maximized" (*Leimbach et al. 2017, 32*) as the sum of welfare across different regions.

remains a substantial value judgment that is surprisingly underappreciated. Doo-ley et al. observe: "Quantified approaches also often implicitly assume that cost optimization is neutral, requiring no ethical justification" (*Dooley et al. 2021, 301*). IAM studies on feasibility often do not mention the CEA framework of the scenario evidence they use (cf. *Gambhir et al. 2017*) or do not discuss its influ-ence further (cf. *Loftus et al. 2015*). The AR6 uses the term "cost-minimizing pathway," but it does not extensively discuss value implications.

From an ethical perspective, CEA can be viewed as implementing welfare economics or, more generally, discounted utilitarianism. As such, it poses a clear value commitment, which competes with *deontological principles* or considera-tions from *virtue ethics*.[6] The apparent innocuousness of CEA seems obscure in this light, but it might be explainable by what we can call "commonsense effi-ciency." Commonsense efficiency is the case "of an individual's taking the least costly, effective means to achieving some particular end" (*Buchanan 1985, 8*). Commonsense efficiency seems to be an uncontroversial value commitment, as it seems always better to reach a given end with fewer costs when possible. How-ever, on closer look this is only uncontroversial as long as all costs fall on the same agent and there are no opposing moral obligations. Both are not the case in mitigation modeling. IAMs distribute mitigation burdens across regions and generations, and a range of moral considerations apply in this context. Therefore, cost-efficiency in IAMs is far from uncontroversial as a value judgment.

There are two ways, how economists defend the neutrality of cost-efficiency. The first is to take efficiency as an idealized *description* of how free markets operate. This approach is familiar from the debate on positive and normative economics (cf. *van Laar and Peil 2009*). Nordhaus, for example, writes:

> "the use of optimization can be interpreted in two ways: they can be seen both, from a positive point of view, as a means of simulating the behavior of a system of competi-tive markets and, from a normative point of view, as a possible approach to comparing the impact of alternative paths or policies on economic welfare" (*Nordhaus 2013, 1081*).

---

[6] *Deontological theories* argue for ethical principles that we must conform to, regardless of the concrete consequences of the individual acts. Most deontological theories are inspired by the ethics of Immanuel Kant (*Kant 1995, 1786 [2007]*), for instance, contractualist theories (cf. *Rawls 1999*; *T. M. Scanlon 1998*). Classic *utilitarianism* goes back to Jeremy Ben-tham (*1907 [1789]*) and John Stuart Mill (*1861*). In its most basic form, it argues that all human action should maximize utility as the sum of the well-being of all beings. *Virtue ethics* grounds ethical justification in the morally good character traits and delineates important values and how they can guide us (cf. *Aristoteles 1985*).

Nordhaus argues that we should choose the parameters of the welfare function to represent "the outcome of market and policy factors as they currently exist" (*Nordhaus 2013, 1081*). If this is done, economists do "not make any case for the social desirability of the distribution of incomes over space or time of existing conditions, any more than a marine biologist makes a moral judgment on the equity of the eating habits of whales or jellyfish" (*Nordhaus 2013, 1082*). In this positive interpretation "we can interpret optimization models as a device for estimating the equilibrium of a market economy" (*Nordhaus 2013, 1111*).

However, such reasoning is problematic for several reasons. Not only deviate real-world economies in many ways from the idealized markets of economists, not least due to climate change being the "greatest market failure the world has ever seen" (*N. Stern 2007*, viii; cf. *Trutnevyte 2016*). More importantly, IAMs use efficiency in effect as a criterion for the distribution of mitigation burdens imposed by the global carbon budget. How to distribute these burdens poses *a real normative choice*. As Dooley et al. (*2021*) write, "there is no ethically neutral position in the climate context, pretending to be value-free obscures unconscious biases under a veneer of neutrality, particularly in quantitative modelling" (*Dooley et al. 2021, 304*). We will be concerned with this argument again in other sections.

The second way economists could hold on to the neutrality of cost-efficiency in IAMs is by pointing out that modeling results must be strictly separated from their practical implementation. This claim of *separability* is central to the self-understanding of modelers. For instance, the IPCC AR5 writes: "Regional IAM results need thus to be assessed with care, considering that emissions reductions are happening where it is most cost-effective, which needs to be separated from the fact who is ultimately paying for the mitigation costs" (*Riahi et al. 2022, 13*). Mainstream economics assume that all goods are exchangeable and that costs can ultimately be redistributed separably from who is implementing the mitigation itself. Therefore, efficiency gains can be used to compensate people with higher costs. Given this compensation logic, "inefficiency is pure waste; it does no one any good" (*Broome 2012, 40*).

This would seemingly allow IAM results to circumvent value judgments, as the value questions only appear later when deciding who should pay for the efforts necessary in a given scenario. However, *first*, such redistributions are highly unlikely to occur. The latest IPCC AR6 estimates that an "equitable emission trading scheme would require very large international financial transfers, in the order of several hundred billion USD per year" (*Riahi et al. 2022, 88*). Implementing such transfers seems unrealistic given the world's power structures, especially as "transfers of anything near this magnitude are not under discussion as part of any climate-policy package" (*Budolfson et al. 2021, 830*). Making modeling results conditional on this assumption compromises their relevance in guiding international climate policymaking and conflicts with the acclaimed goal of IAMs to be policy-relevant. *Second*, the underlying logic of perfect compensation is in itself a value commitment, as I will explain in Sect. 6.4.

Third, regional and disaggregated results become increasingly influential, especially in the shift towards assessing feasibility (cf. *Brutschin et al. 2021*; *Vinichenko, Cherp, and Jewell 2021*). Data on concrete indicators for feasibility, such as overall costs, investments, or stranded investments in fossil infrastructure, are a direct product influenced by the CEA framework. Moreover, regional results are directly affected by cost-effectiveness, as mitigation efforts in the models occur where they produce the least overall costs. van de Ven et al. (*2023*) calls the use of regional results a "key novelty" of recent frameworks, "allowing us to assess to what extent and from which perspective countries' policy targets, NDCs, and LTTs are feasible" (*van de Ven et al. 2023, 571*). "Many policy discussions have been guided by IAM-based quantifications, such as the required emission reduction rates, net zero years, or technology deployment rates required to meet certain climate outcomes," Riahi et al. (*2022, 13*) write.

Separability should thus be rejected in this context, as it cannot relieve the cost-effectiveness framework from being an influential and contested value choice in IAMs. As it stands, it, moreover, represents a dubious value judgment from a perspective of justice. Efficiency disregards important considerations of justice, such as historical responsibility, different capabilities, and the ethical necessity to meet basic needs. Such considerations are not only central to the justice debate but also a recognized part of the UNFCCC charter, which states to "protect the climate system [...] on the basis of equity and in accordance with their common but differentiated responsibilities and respective capabilities" (*UNFCCC 1992*, Article 3.1).[7] Efficiency , in contrast, is only a secondary consideration in the charter behind such implicit justice consideration. Efficiency, however, is central to how the models operate. It may be a defensible criterion *as long as* other ethical considerations are sufficiently considered. A promising way forward is, thus, *to confine the influence* of efficiency as a value judgment in the relevant respects. One important step in this direction would be to reject efficiency in relation to global burden-sharing in favor of explicitly modeling different burden-sharing principles (cf. *Budolfson et al. 2021*). Efficiency is one of several ethical principles that can

---

[7] IAMs include voluntary pledges made by countries. However, for countries without a pledge on emission reduction, studies such as Gambhir et al. (*2017*) assumed simply no specific cap: "In many cases, it makes most sense to simply not impose a cap on regions representing these countries—or combinations of these countries—in the TIAM, WITCH, and MESSAGE models" (*Gambhir et al. 2017*, Appendix A).

be implemented in modeling exercises. What constitutes a just distribution of mitigation burdens is a vital ethical question and has seen a fair share of ethical literature, which provides robust and concrete mid-level principles (cf. *Shue 1999*; *Caney 2005, 2018*; *Vanderheiden 2008*; *Page 2008*; *Singer 2010*; *Schüssler 2011*).

Dooley et al. (*2021*) provide an especially condensed discussion of candidate principles, highlighting various ethical aspects such as historical responsibility, different capacities, subsistence, and other important dimensions of justice that are relevant to it. Holz, Kartha, and Athanasiou (*2018*) also argue for separating ethical choices from the quantitative "equity modeling," as they call it. Fyson et al. (*2020*) apply explicit burden sharing to the issue of CDR, concluding that "fair-share outcomes for the United States, the European Union and China could imply 2–3 times larger CDR responsibilities this century compared with a global least-cost approach" (*Fyson et al. 2020, 1*). Chen et al. (*2021*) modeled an equal per capita distribution of the Paris budget, showing significant differences in regional fossil reliance and energy investments between the principles. Discussions of such principles can also be found in Du Robiou Pont et al. (*2017*) but need to be applied to IAMs more directly.

To sum up, much of the scenario evidence from IAMs is conditional on the assumption of Cost-Effectiveness. This represents an influential value judgment embedded in IAMs. Arguments for its neutrality fail, as efficiency-based modeling cannot be viewed as descriptive in this context, nor can it be separated from the normative question of burden sharing. Implicitly relying on efficiency-based modeling results without discussing its value dimension risks highly unfair burden-sharing principles being embedded in feasibility assessments.

6.4   The Concept of Well-being

The previous sections showed the value judgments implied by the cost-effectiveness framework of IAMs. This section investigates a closely related value judgment in the definitions of *well-being*. Currently, most IAM evidence relies on understanding well-being in terms of consumption. This is a particularly narrow form of valuation. Objective-list conceptions of well-being can provide an alternative value outlook for IAMs, as they allow for the representation of basic needs and requirements of people more directly in IAMs.

IAMs rely in some way or another on a Social Welfare Function (SWF), which aggregates individual well-being into social welfare. This function is used to assess what costs particular mitigation strategies involve. The concept of well-being defines what counts as costs and, thus, influences what kind of pathways are determined in IAMs. Philosophically, well-being is what is ultimately good for a person (*Reiss 2013, 214*).[8] There are three strands of philosophical theories on well-being (*Parfit 1984*; *Reiss 2013*; *Crisp 2023*): (1) *hedonism*, (2) *desire theories*, and (3) *objective list theories*. Classic utilitarianism is based on *Hedonism*, which in its simplest form takes what is good for a person to be ultimately the presence of pleasure and the absence of pain (*Bentham 1907 [1789]*; cf. *Mill 1861*). *Desire Theories* depart from the experiential understanding of well-being and replace it with the satisfaction of wants or, as economists would put it, preferences. In many cases, what we desire is closely related to what is pleasurable. However, things are sometimes good for a person, though they do not necessarily translate to pleasure, even in the long run. The third approach, *Objective List Theories*, departs from the subjective perspective by proposing an objective understanding of what constitutes a good life (cf. *Aristoteles 1985*). For instance, one might understand well-being in terms of a list of criteria such as having bodily integrity, having the basis for social recognition, being able to play, experiencing a full emotional life, and so on (cf. *Nussbaum 2012*).

Thus, welfare economics and integrated modeling are based on a specific version of the desire theory. Well-being is taken to consist of the fulfillment of people's actual wants. Economists assume that people reveal their preferences in their market behavior, thus allowing for its empirical quantification (cf. *Hausman and McPherson 2006, 19*; *Beckerman 2017, 8*; *Crisp 2023*). Economists conceptualize this by assuming a homogenous good they call "consumption." As Nordhaus (*2013, 1083*) writes, this good should "be viewed broadly to include not only food and shelter, but also non-market environmental amenities and services." As an abstract, homogenous good, it is meant to cover everything valuable to people.

Conceptualizing well-being as "consumption" is a substantive value judgment. *First*, not all valuable things are traded on the market. For instance, many ecological goods, or goods such as the value of political stability, would be missed

---

[8] There have been some philosophical doubts on whether such a "master value" is a plausible concept at all (*T. M. Scanlon 1998, 108–43*; *Moore 1903 [1993]*), which, if true, would be a general problem for the welfare approach.

in the economic valuation (*Sterner and Persson 2008*).[9] There is also a feminist perspective to this, as certain goods, like child-rearing and household work, never enter the market but are performed overwhelmingly by women.[10] How well such non-market goods translate into consumption is an open question (cf. *Atkinson, Bateman, and Mourato 2014*; *Mintz-Woo 2021b, 524*). *Second*, impoverished households often have no measurable income and thus do not appear on the market at all. Welfare optimization might completely neglect the losses in the welfare of the very poor. Weyant writes on CB-IAMs: "This means that if an optimal carbon tax is computed for a market or country based on global costs and benefits, the impacts on the world's poorest people will not be included" (*Weyant 2017, 126*). In IAMs, this would exclude mitigation costs occurring to the poorest household—a highly dubious ignorance in the policy-relevant projection of various global futures.

Integrated modeling, as it stands, is based on a relatively narrow form of valuation (cf. *Gardiner 2011, 252*) that neglect critical ethical issues in determining mitigation pathways, including representing the most vulnerable parts of the population. Like the framework of efficiency, economists sometimes consider the proxy of consumption to be neutral, as it takes people's preferences as a given. Moreover, it has received comparably little discussion, perhaps because it is central and fundamental to the economic lens. Lamb and Steinberger (*2017*) observe that well-being receives "relatively little attention in comparison to the economic and technical features of mitigation" (*Lamb and Steinberger 2017, 11*).

Objective list theories provide a good and fruitful alternative to this. They can solve the representation deficit regarding people who are very poor and come with

---

[9] We can distinguish between *weak* and *strong sustainability* in environmental economics. *Strong sustainability* holds that there can be no substitution for losses of some goods in terms of consumption. *Weak sustainability* assumes that substitution is possible and sustainability only demands that future generations are not made worse overall without necessitating specific goods' sustainability. Gardiner (*2011, 263*) argues against *weak sustainability* with his "dome scenario." In this scenario, future generations live under an artificial dome due to a degraded environment but have higher welfare in terms of consumption. Standard economic analysis suggests this is preferable if consumption in these domes is high enough. Gardiner argues this is a "highly contentious position, and one embodying a major value assumption" (*Gardiner 2011, 263*). The precautionary approach of PB-IAMs is a case of strong sustainability concerning climate change but not concerning other (environmental) goods. The survey of van Soest et al. (*2019*) on the state of SDG representation in IAMs also shows that only a few studies have analyzed the biodiversity impact of land-based mitigation, a topic "deemed as important according to the expert survey" (*van Soest et al. 2019, 214*).

[10] A.C. Pigou, for example, pointed out long ago that if he were to marry his housekeeper, the national GDP would decrease (*Beckerman 2017, 12*).

a solid basis in moral theory. An implementation of this approach would be to start with a conception of basic human requirements and explicitly represent such aspects of well-being in the models. Philosophically, discussions on needs and requirements for a good life have been developed, for instance, in the capabilities approach, which proposes some crucial elements of a minimally decent life (cf. *Sen 1984*; *Neef-Max 1991*; *Doyal and Gough 1991*; *Nussbaum 2012*). In a recent paper, Rao and Min (*2018*) rely on this philosophical background in defining the "Decent Living Standard" (DLS). These are "a set of material requirements that are essential for human flourishing" (*Rao and Min 2018, 226*). These material needs represent the basic entitlements of all humans. They write that DLS should "guide the establishment of reference budgets and living wages, and development policies" (*Rao and Min 2018, 242*).

Modeling objective aspects of well-being and prioritizing meeting basic needs can lead to very different mitigation strategies and measures for staying within low-temperature goals, almost entirely missing from the scenario evidence from IAMs at the moment. In a widely recognized study, Kikstra et al. (*2021*), for instance, uses a simple model to estimate the energy needs for achieving decent living standards across the globe. They find that "equity in living standards demands significant convergence between rich and poor countries' energy use" (*Kikstra et al. 2021, 10*). Similar studies also find that "unprecedented reductions in income and energy inequalities are likely to be necessary to simultaneously secure a climate-safe future and decent living standards for all" (*Millward-Hopkins and Oswald 2023, e147*). These findings starkly contrast with the results of consumption-based modeling, which currently almost ubiquitously show significantly higher energy consumption in the Global North than in the Global South (*Hickel and Slamersak 2022, e629*). Scenarios in which energy consumption equalizes are absent from the scenario space of IAMs. Kikstra et al. suggest "distinguish basic energy needs within emissions pathways" as a next step for integrated modeling (*Kikstra et al. 2021, 9*).

To conclude, a significant value judgment in IAMs concerns the definition of well-being. Well-being provides a particularly approachable lens to shed light on more aspects of climate mitigation and challenge the established paradigm. Most existing scenario evidence is based on consumption as a measure of well-being, which faces a range of objections from a justice perspective. An alternative value perspective could be brought in by representing objective-list theories in mitigation scenarios, for instance, by using a conception of basic requirements for a good life.

6.5   Representation of Social and Global Inequality

After discussing the general framework of cost-effectiveness and well-being, this section explicates another value judgment within the Social Welfare Function (SWF). The SWF aggregates individual well-being into social welfare. Different forms of aggregation are available for this purpose, which will diverge in how they treat differences in well-being between individuals and groups. The representation of social and global inequality within IAMs is a value judgment, and most scenario evidence currently disregards inequality as an explicit consideration. This section will briefly present two other forms of aggregation, which provide readily available alternatives to shed light on global and social inequality. Whatever aggregation one assumes, one makes a value choice.

The dominant form of SWF in IAMs is that of *discounted utilitarianism.* According to it, social welfare is the unweighted sum of each person's well-being. Differences in well-being between people do not affect the overall level of welfare. As explained in the last section, well-being is mainly understood as a function of consumption levels. In translating consumption levels into well-being, however, inequality in consumption plays a role. Economists generally assume that consumption has a marginally declining utility, and thus, they multiply consumption levels with a parameter called *"elasticity of marginal utility,"* or η.[11] η describes the degree to which an additional consumption unit is less valuable to a rich person than a poorer person.

Usually, this would imply that η governs how mitigation burdens are distributed across people with different consumption levels. Since IAMs minimize costs, η would represent a normative parameter that determines how the models distribute mitigation burdens between different social and global groups with different starting income levels. (η is also integral to so-called growth discounting, which will be the topic of the next section.[12]) Higher values for η shift burdens to more affluent social groups, whereas a lower η will distribute burdens more equally. Thus, η would typically be a normative parameter for the "aversion to inequality," which we build into the model across time, regions, and society.

However, the influence of η as a parameter on inequality aversion is blocked by two other aspects common in IAMs. Concerning social inequality, IAMs

---

[11] For instance, the logarithmic utility function in the ReMIND model represents an elasticity of marginal consumption η of 1 (*Luderer et al. 2015, 11*)—the same value as applied in N. Stern (*2007*) and Nordhaus (*2013*).

[12] As we will see in the section on discounting, η has multiple normative roles. It can also be seen as a parameter for our attitude towards risk in the intertemporal dimension (*Mintz-Woo 2021b*).

typically only assume a single "representative household" per region for simplicity reasons. Common SWFs multiply the consumption of an average person
with the population number to receive social welfare, for instance, done this
way in the ReMIND model (*Luderer et al. 2015, 10*). Therefore, differences in
consumption between different social groups do not appear in the model and distributive aspects concerning social inequality are typically neglected in mitigation
pathways. As different measures to mitigate emissions will affect social groups
unequally, this is a limitation of IAMs that could be overcome by improving
household representation but would also require representing mitigation policies
in greater detail, going beyond the paradigm of a uniform carbon price (*Rao
et al. 2017, 860*). Such distributional issues are a significant concern, both in the
climate justice literature and to policymakers, and have proven to have a significant influence on crucial results in Cost-Benefit models (cf. *Anthoff and Tol 2010*;
Budolfson et al. 2017) and bottom-up studies (cf. *Daioglou, van Ruijven, and van
Vuuren 2012*). As it stands, however, η does not address social inequality.

On the global scale, η is also blocked from accounting for distributive issues.
IAMs divide the world into different regions with different levels of wealth. The
influence of η on the distribution of mitigation burdens is nevertheless blocked, as
interregional redistribution is prevented in optimizing models, such as ReMIND,
by the use of so-called *Negishi weights* (*Luderer et al. 2015, 11*; *Leimbach et al.
2017, 32*). This needs some explanation. Normally, positive values of η would
lead multi-regional IAMs to *redistribute* wealth between the different regions.
As regions have different starting consumption levels, η implies that any transfer
of financial resources from a lower-income to a higher-income country would
increase total welfare within the models. As IAMs maximize global welfare (or
minimize costs in terms of welfare "losses"), they have an inherent redistributive
tendency. However, this tendency is seen as a bug rather than as a feature of
mitigation pathways:

> "Another objection is that certain consequences of utilitarian social choice rules are
> implausible. For example, the commonly made assumption of diminishing marginal
> utility of income implies that a major redistribution of income would be required
> between developed and developing countries under a utilitarian social welfare func-
> tion. This may not reflect actual social preferences since this does not match actual
> government policy" (*Botzen and van den Bergh 2014, 9*).

As large-scale redistributions are considered politically infeasible, and as it is considered beneficial to separate the issues of global inequality and climate action,
modelers use Negishi weights to block this redistributive tendency. Negishi
weights are attached to the regional utility gains and adjusted to equalize marginal

utility gains across regions. This way increases in consumption are treated equally in all regions *regardless* of the starting income level (*Stanton 2009, 7*).

Whether the underlying feasibility judgments have epistemic merits is an open question. Such redistributions would be clearly feasible, even if they are not likely to occur. Moreover, it is widely accepted that climate policy will involve some transfer from historically high emitting countries to low emissions countries. The COP 2009 in Copenhagen, for instance, included a pledge by rich nations to transfer a total of US$100 billion yearly from 2020 onwards to poorer nations to help them adapt to climate impacts and mitigate their emissions.[13] More importantly, though, Negishi weights represent a significant value commitment in IAMs:

> "Negishi weights freeze the current distribution of income between world regions; without this constraint, IAMs that maximize global welfare would recommend an equalization of income across regions as part of their policy advice. With Negishi weights in place, these models instead recommend a course of action that would be optimal only in a world in which global income redistribution cannot and will not take place" (*Stanton 2009, 2*).

IAM results are conditional on fixing income levels between regions. In effect, welfare-optimizing IAMs, which use Negishi weights, actually prioritize well-being in rich regions compared to poor regions. IAMs "are acting as if human welfare is more valuable in the richer parts of the world," as Stanton, Ackerman, and Kartha (*2009, 176*) put it.[14] Mitigation pathways thus opaquely support the continued economic dominance of the Global North over the Global South and disallow equalization of global income. This is a highly dubious ethical position to assume in light of existing theories of global justice (cf. *Beitz 2001*; *Moellendorf 2011*). Negishi weights imply that η does not represent the inequality aversion between different regions.

However, η is not the only way to address inequality. Conceptually, η is actually a statement about the *personal* value of consumption, not its *social* value. Scholars have thus suggested representing questions of distributive justice by

---

[13] This pledge has so far not been met by developed nations (*Timperley 20.10.2021*), but it shows that any COP meeting needs to involve talks on financial transfers.

[14] Stanton gives a second reason to reject Negishi weights, independent of its value choice. As applied in IAMs, diminishing marginal utility is allowed to influence the intertemporal distribution of welfare by discounting the consumption of (wealthier) future people. It is, however, not allowed to influence the distribution of wealth between regions. Populations distanced in time are treated differently than populations distanced in space. This is an inconsistency in IAMs that cries for explanation (*Stanton 2009, 11*). Sterner and Persson (*2008, 66*) also notes that using distributional weights was the norm in the economics of the 70 s.

using other explicit forms of welfare functions. I will briefly discuss two alternatives to utilitarian SWFs in prioritarianism and sufficientarianism. While there are many ways to think about social welfare, these two represent proximate alternatives that allow us to investigate value questions regarding inequality.

The first alternative is a prioritarian SWF (*Adler and Treich 2015*; *Adler 2016*, *2019*). The idea of Prioritarianism is to assign people who are worse off a priority in distributing welfare. In IAMs, this would imply that poorer people are given a priority over richer people in terms of welfare, when deciding who should bear the burdens of mitigating climate change. It is different from the marginal utility of consumption described above, as the prioritarian parameters apply directly to well-being, not consumption (*Lumer 2005*; *Adler 2011*).[15] Prioritarian SWFs attach a weight to equality and, thus, have an egalitarian tendency. Different philosophical justifications for egalitarianism are available in the literature. Egalitarianism might be justified as an approximation towards what luck-egalitarians see as a just distribution, namely a society that takes inequality in resources to be justified only when it is the result of people's choices, not when it is the consequence of arbitrary factors such as natural talents, circumstances, or pure luck (cf. *Rawls 1999*; *Dworkin 1981*). It might also be defended on instrumental grounds, as large inequalities in welfare might harm equality in political goods, such as having an equal say (cf. *Piketty 2014, 2015*; *T. Scanlon 2018*).

Such a valuing of equality in the SWF would, of course, pose an explicit value choice. For this reason, economists such as Partha Dasgupta have criticized it as ad-hoc:

> "Some ethicists have proposed an ethical theory they call 'prioritarianism,' which says that an increase in the well-being of a rich person [...] should be assigned less social value than the same increase in the well-being of a poor person [...]. I have not understood why such an ad hoc ethical principle should be awarded a name" (*Dasgupta 2008, 146 fn.4*; cited after *Adler and Treich 2015, 286*)

---

[15] It assumes that a transfer *in well-being* from a richer to a poorer person would make societies better off, even if the overall sum of well-being stays the same. Prioritarianism operates on the level of *aggregating* well-being across individuals. In economics, this is known as the Pigou–Dalton principle. It states that a "pure, non-rank-switching transfer of well-being from someone better off to someone worse off, leaving everyone else unaffected, is a moral improvement" (*Adler and Holtug 2019, 4*)

However, relying on a utilitarian SWF would not relieve IAMs from making this value choice, as it is hard to see why disregarding differences in well-being is not an ethical judgment.[16]

A third alternative that merits serious consideration is *sufficientarianism* (cf. *Crisp 2003*; *Gosseries 2016*). It could be combined with traditional conceptions of well-being as consumption. However, it most naturally connects to the discussions above of objective-list conceptions of well-being. Once we differentiate the basic requirements of a decent life within IAMs, we assume these requirements must be met first. Nussbaum (*2012*) follows such an approach. Many ethicists agree that guaranteeing every person sufficient resources to lead a decent life has moral priority over welfare motives. This is the basic premise of sufficientarianism (*Fourie 2016*). Since different mitigation strategies have dramatically different implications regarding basic requirements for a decent life, for instance, concerning poverty, food security, clean water, and energy access, relying on sufficientarian versus utilitarian aggregation will make a big difference in IAM pathways.

So far, this has been done only in bottom-up studies. D. W. O'Neill et al. (*2018*), for instance, investigates how the "provisioning systems," the physical and social systems, can meet basic needs in terms of nutrition, sanitation, income, access to energy, education, social support, equality, democratic quality, and employment (*D. W. O'Neill et al. 2018, 89–90*). Including sufficientarian mitigation pathways would be valuable. The scenario evidence from IAMs and subsequent feasibility assessments does not represent this value outlook. For instance, of the 230 scenarios compatible with 1.5 °C in the AR6, only two scenarios target sufficiency measures. However, as Yamina Saheb explains in an interview, both "didn't make it to the IPCC database, because to be able to submit to the database you need to have resources—especially human resources. Realistic scenarios for a liveable planet like the ones developed by Kai [Kuhnhenn et al.] and Julia [Steinberger et al.] are often developed by very small teams, sometimes on a voluntary basis, putting in lots of hours during their free time" (*Saheb, Kuhnhenn, and Schumacher 08.06.2022*).

---

[16] Adler and Holtug (*2019, 13*) also gives this line of response. They write: "Utilitarianism, best understood, does not deny the distinction between the well-being goodness of life and its moral goodness. Since both academic philosophers and lay-persons discuss the nature of well-being independently of moral debates, denying this distinction would be very problematic. Rather, the utilitarian should concede the distinction but argue on substantive moral grounds for the formula $\sum_{w_i}$" (*Adler and Holtug 2019, 4*).

Finally, the combination of cost-minimization, well-being, and the form of aggregation plays out in the reliance on growth as a background of most scenario evidence from IAMs. IAMs currently almost exclusively rely on growth-based strategies to mitigate climate change. Almost all IAM research assumes a steadily increasing consumption for the future. Keyßer and Lenzen (*2021*) observe that "[n]one of the 222 scenarios in the IPCC SR1.5 and none of the shared socioeconomic pathways projects a declining GDP trajectory" (*Keyßer and Lenzen 2021, 2*). While degrowth is a vital part of the public discourse, it is not represented within the feasibility assessments of IAMs at all. Researchers have argued to develop more IAM scenarios that do not depend on economic growth and, further, that non-IAM scenarios should also be included in the database for mitigation scenarios (cf. *Gambhir, Ganguly, and Mittal 2022*) to compensate for the bias in IAM research.

To summarize, I argued that whatever form of aggregation in SWF one chooses, one makes an ethical choice. IAMs use η as a parameter to account for the marginal declining utility of consumption. However, this parameter does not adequately account for inequality since the lack of household representation and Negishi weights block its influence on issues of inequality. How to deal with existing inequalities in climate mitigation strategies is a significant value judgment. Prioritarian and sufficietarian SWF are viable alternatives that should be included in the projection of IAMs.

6.6 Intergenerational Burden Sharing and Discounting

IAM projections span the whole of the 21st century. Different mitigation strategies will affect generations unequally, so IAMs need a method for comparing costs and burdens across time. Economists use the *social discount rate (SDR)* for this purpose. The SDR translates future costs into a "net present value," making it possible to determine cost-effective strategies across time. Since IAMs minimize overall cost, higher discount rates decisively backload mitigation pathways, shifting mitigation burdens to the future.

Discounting has already drawn intense interest, both in economics and ethics. Some call the discount rate a "success story" of value transparency (*Bistline,*

Budolfson, and Francis 2021, 3; cf. *Carrier 2021*).[17] However, this has yet to fully transfer to PB-IAMs, where high discount rates influence mitigation pathways in the form of a constant background parameter. This section will explicate the value choices involved in discounting and their influence. It will argue against "descriptivism" within the context of IAMs, provide alternative value choices, and highlight ways to limit the influence of discounting.

Discount rates in IAMs apply directly to welfare, and given the cost minimization, they affect *when*, *where*, and *with which means* climate change is mitigated in the models. Due to the compounding nature of discounting, the chosen SDR substantially influences what kind of projections the models produce. High discount rates make future efforts appear comparably cheap within the models, thus shifting much of the mitigation burdens to the future. While the actual numbers for the SDR appear small, its influence is substantial. At an SDR of 1%, Mitigation costs of $1.000 that arise in 30 years equal to present costs of $741. At an SDR of 3%, this number goes down to $411; at 5%, it is only a third at $231. Most mitigation pathways from IAMs rely on an SDR in the range of 4–6%, including the scenario evidence produced for the IPCC report (*Clarke et al. 2014*; *Masson-Delmotte et al. 2018*; *Riahi et al. 2021*; *IAMC 2022*) and the discussed feasibility assessments (*Gambhir et al. 2017*; *Brutschin et al. 2021*; *Vinichenko, Cherp, and Jewell 2021*).[18]

Discount rates reflect how we should distribute mitigation burdens across time and how much risk we can leave to future generations. This is an ethical question, and answers to them become ingrained in choices of the discount rate. "[D]etermining the appropriate social discount rate is mostly a normative problem" (*Kolstad et al. 2014, 229*), the IPCC wrote in its "ethics chapter" in the AR5 back in 2014.

IAMs, however, largely rely on a methodology known as "descriptivism," which argues to determine the SDR so that it reflects "real" interest rates observable in the market: "values of discount rates adopted in DP IAMs are around 5%–6% per year […], in line with market interest rates" (*Emmerling et al. 2019, 2*). The ReMIND documentation states that the SDR is determined to be

---

[17] The dispute between Nordhaus and Stern received much attention, in which discount rates of 6 and 1.7% (among other factors) led to highly disparate outcomes (*N. Stern 2007*; *Nordhaus 2007*). Cf. arguments in Parfit (*1984, 480–86*), Broome (*1994*), Nordhaus (*1997*), Dasgupta (*2008*),Caney (*2009*), Gardiner (*2011, 270–98*), Moellendorf (*2013*), and Mintz-Woo (*2021a*). An overview of the debate is given by Dennig (*2018*).

[18] The AR6 reports modeling results with a discount rate of 3%, but this rate is applied only ex-post to the modeling results. The models mostly use higher internal rates, and thus, mitigation pathways are shaped by these higher SDRs.

"in line with the interest rates typically observed on capital markets" (*IAMC 2022*).[19] *Descriptivism* tries to shortcut this normative question by relying on market rates as an empirical base. The rationale behind this is that market rates are seen as revealing people's preferences and thus could be considered neutral or a seemingly democratic choice. However, as has been argued extensively in the literature, this approach is inadequate in this context (cf. *Broome 1994*; *Roser 2009*; *Fleurbaey and Zuber 2012*; *Moellendorf 2013*). Future generations do not participate in the market, but their welfare is most seriously affected by discounting (*Broome 1994*). More importantly, observed preferences are methodologically insufficient for answering the *normative* question of how we should value future generations' welfare and, consequently, how to distribute burdens arising from mitigating climate change.[20] Descriptivism is one possible answer to this question, but it cannot shortcut its normative quality (cf. *Gardiner 2011*).[21]

If we accept that the SDR is a value-laden parameter, we need to ask what value judgments are involved in determining the SDR one way or the other. We, therefore, need to dive into some more specifics concerning discounting. The basic thought behind discounting is that a dollar now is more valuable than a dollar in the future because *(a)* welfare in the future itself is less important than welfare now, and *(b)* we expect to be more prosperous in the future. Thus, we will care less about extra costs in the future. These two rationales are represented by the two elements in the Ramsey rule, the widely accepted framework for discounting in welfare economics (*Ramsey 1928*), given by the formula $SDR = \delta + \eta * g$.

---

[19] Mintz-Woo calls it an "investment-based" approach or "opportunity cost discounting" (*Mintz-Woo 2021a, 94*), as its fundamental assumption is that we should view climate change as an investment problem for which we need to keep the "opportunity cost of capital" (*Nordhaus 2007, 689*) in mind and, thus, should set discount rates to the "real returns" in comparable markets (cf. *Posner and Weisbach 2010*, ch. 7).

[20] In other words, observing individual attitudes is simply a different subject matter. Measurements on individual time preferences would not help much either by themselves since they most likely will vary significantly across different contexts (*Mintz-Woo 2021a, 101–2*).

[21] There is also a debate within Descriptivism whether the chosen SDRs are adequate. Dasgupta (*2008*) argues, for instance, that observable rates of return on investments are likely too high since the negative externality of climate change distorts the observable prices. Moreover, even if we use the "real" interest rates, some discount rates in IAMs appear relatively high. For example, an expert survey conducted by Drupp et al. (*2018*) among 200 experts on social discounting returned a median value of 2% for the "real risk-free interest rate." The median value for their best guess on the appropriate overall social discount rate was 2 percent, informed by descriptive and normative considerations. 92% of responses locate the SDR in "the interval of 1 to 3 percent" (*Drupp et al. 2018, 111*). Thus, even if we accept this approach for IAMs, there are good reasons to reject the current high rates.

The first summand of the Ramsey rule (δ) is known as *pure time discounting*. δ is applied to welfare directly, and thus, it has direct normative significance. It represents the value judgment of how much we value welfare or utility occurring at different times. It represents the degree to which we disregard future people's well-being in our long-term mitigation strategies. Positive δ are sometimes defended by pointing to social impatience, that is, people's preference to enjoy additional consumption now instead of later (cf. *Dasgupta 2008, 145*). Mostly, though, contributions from economists and ethicists agree that δ should be equal or very close to 0% (cf. *Dasgupta 2008, 157*; *Broome 1994, 131*; *Caney 2014a*; *Parfit 1984, 480–86*). Ramsey himself described a positive rate of pure time preference to be "ethically indefensible" and arising "merely from weakness of the imagination" (*Ramsey 1928, 453*).[22] This is in stark contrast to many existing IAM studies, which, based on the descriptivist approach, deploy positive values for δ.

The most basic argument for $\delta \approx 0$ is that the value of a person's well-being cannot depend on such partial consideration as the date they are born. Employing a positive time preference would do exactly this. Pure time discounting at 3% would be equivalent to assuming that the welfare of a person "born in 1960 should 'count' for roughly twice as someone born in 1985" (*Dietz, Hepburn, and Stern 2009, 13*; *Dasgupta 2008, 157*). This is by many seen as implausible and ethically indefensible. Thus, there are strong reasons for a pure time discount rate of zero. Determining a positive δ based on current people's preferences would hardcode our generation's selfishness into the models.[23]

---

[22] Possible (ethical) reasons for a positive δ are the following, which may be used to defend current pure discount rates if made explicit: *(1)* A non-zero chance of extinction could justify pure time discounting, though only very small values appear credible, e.g., 0.1% in N. Stern (*2007, 31*). *(2)* A special kinship to our contemporaries could justify discounting future people's well-being (cf. *De-Shalit 1995*; *Mogensen 2022*). *(3)* It is argued that ethics would be overdemanding if we set δ to zero (*Mintz-Woo 2021b*). Undiscounted SWF could imply extremely low consumption levels for the present generation, as any investment for the future could have greatly increasing returns. Such a "sacrificing the present for the future" (*Adler and Treich 2015, 283*; *Weitzman 2007*) would be a counterintuitive moral result. However, this argument might be seen as a reason against utility maximization in general (*Roser and Seidel 2015, 52*). Since there is a second component to discounting, this might guard against such outcomes. *(4)* Another (rather ad-hoc) reason is that undiscounted utilitarianism makes the algorithm unsolvable: "This, indeed, is why many economists today use discounted-utilitarianism: not because it has a sound ethical foundation—at least for the discount rates commonly employed—but because it *gives a unique answer* to the problem" (*Roemer 2011, 375*).

[23] As an ethical parameter, we might even conclude that δ should be negative if we think that the current generation has a special duty to mitigate climate change and bear much of the

The second summand of the Ramsey rule ($\eta*g$) is *growth discounting*, which is more complicated yet, in practice, even more influential. Growth discounting occurs because IAMs assume that humanity will be more prosperous in the future. The Stern Review, for example, takes the average GDP to grow from $7.600 today to $94.000 in 2200 (*Sterner and Persson 2008, 67*). Growth leads to positive discount rates due to the already familiar parameter η, the elasticity of marginal utility of consumption.[24] As described in the section on inequality, η reflects the widespread economic assumption that extra consumption has a declining value for a person. If we assume that future generations will, on average, be more affluent, costs occurring in the future would be less burdensome. η can be understood as a measure of the "aversion to inequality" that we want to apply to our assessments with IAMs, here now in the comparison across time and between generations (*Kolstad et al. 2014, 230*; *Nordhaus 2008, 60*). A higher η will shift more burdens to the wealthier future generation and away from the comparably poor current generation.

The exact value of growth discounting depends on a mix of value judgments and empirical estimates concerning how the future economy will develop. The previous section described that IAM pathways assume continuous growth in terms of GDP but that this assumption relies decisively on two value judgments in IAMs, namely on thinking of welfare mostly in terms of consumption and largely disregarding differences between social groups. Accounting for these limitations could imply vastly different discount rates. Fleurbaey and Zuber (*2012*) argue that if we include distributive issues and account for the fact that climate impacts may make some portion of the future poorer, then growth discounting could suggest even negative discount rates. If "the present donor is richer than the future beneficiary" (*Fleurbaey and Zuber 2012, 586*), discount rates turn negative even with conventional values for η, something that could materialize in the context of climate change. The future poor will suffer disproportionately from climate impacts and carry much of the unaccounted costs of climate mitigation, for instance, by suffering from water stress and food insecurity due to heavy reliance on bioenergy. Such an IAM scenario will likely imply much higher short-term mitigation requirements than currently suggested by mitigation pathways.

Most modeling studies report discount rates. However, it is often not discussed as a value judgment, and the current literature needs more variation of

---

burdens of mitigation itself. One ethical rationale is that even low-temperature goals come with significant burdens that justify relieving the future to contribute to mitigation.

[24] With typical growth of about 1 to 3% and η being between 0.5 and 4, this second summand is, in general, more influential than the conventional pure time preference rates but has been largely neglected by climate ethics (*Mintz-Woo 2021a, 93*).

this parameter to understand its influence on central modeling results. While the AR6 acknowledges the value problem in its glossary, stating that the "choice of [the] discount rate(s) is debated as it is a judgment based on hidden and/or explicit values" (*van Diemen et al. 2022, 1800*), it does not include any justice debates in any depth. In contrast to the simpler CB-IAMs, variation concerning the discount rate is also lacking: "Most models have a discount rate of 3–5%, though the range of alternatives is larger. Cost-benefit IAMs have had a tradition of exploring the importance of discount rates, but process-based IAMs have generally not" (*van Diemen et al. 2022, 1875*), the IPCC writes. Current scenario evidence from IAMs mainly relies on high discount rates with a significant influence on the results and thus are biased in favor of present generations. The underlying value judgments are buried in mathematics so that most users do not realize their influence.

The comparably high discount rates in the current scenario evidence significantly backload mitigation pathways and drive reliance on Carbon Dioxide Removal in mitigating climate change within the models. Emmerling et al. (*2019*) write that lowering the SDR "significantly improves intergenerational equity" in the pathways (*Emmerling et al. 2019, 5*), leading to emission profiles in which "the mitigation effort is equally distributed across generations, independently of the scenario and carbon budget considered" (*Emmerling et al. 2019, 5*). Lower discount rates increase short-term efforts dramatically in the models. There is up to a six-fold (!) increase in near-term carbon prices in the modeled pathways when the discount rate is lowered from 6% to 1% (*Emmerling et al. 2019, 3*). Lower discount rates reduce the reliance on Carbon Dioxide Removal; for instance, lowering the SDR from 5 to 2% "represents a reduction of about 300 $GtCO_2$ of net negative emissions across the century" (*Emmerling et al. 2019, 4*), the equivalent of 9 years of current emissions.[25] In short, current high discount rates shift considerable mitigation burdens to the future and towards CDR. I will discuss CDR in more detail below.

A recent development would limit the influence of discounting within IAMs if it were more widely adopted. Newer modeling studies explicitly define the amount of CDR in the scenarios, limiting the influence of discounting on this contentious issue. The publication of Rogelj et al. (*2019*) proposed to make the

---

[25] This considerable influence of the chosen discount rates on pathways with unrestricted CDR is a robust finding (*Grant et al. 2021*; *Gambhir and Tavoni 2019, 407*). Riahi et al. (*2021*) find that reducing the discount rate from 5 to 1% would double the necessary emission reductions by 2030 due to less reliance on CDR, but this finding did not make it to the main text. CDR "may be entering the solution space of IAMs for the 'wrong' reasons (discounting) rather than the role they were originally included for (hedging uncertainties)" (*Köberle 2019, 109*).

amount of overshoot an explicit scenario parameter, thereby turning "questions of intergenerational equity into explicit design choices" (*Rogelj et al. 2019, 357*).[26] This has been used in the IPCC AR6. Turning aspects of models driven by high discount rates into explicit scenario parameters is critical for relieving the discount rate from its ethically consequential role.[27] Either way, though, the discount rate remains a value-laden parameter.

To conclude, discounting represents an influential value question in integrated modeling. This section rejected attempts to shortcut its normative content by relying on observed interest rates. Pure time preference is a direct normative parameter reflecting our attitude toward future people's well-being. Growth discounting presents us with a mix of empirical assumptions and value judgments on attitudes toward risk and inequality across generations. Given that IAMs rely on a narrow valuation form in related aspects, one might directly vary the SDR to reflect different plausible value-laden scenarios. Especially lower discount rates need to be added.

6.7   Choice of Domain

The last value aspect relating to welfare is the choice of domain. Whose welfare is considered in modeling different mitigation strategies is an evident value judgment but one that is seldom discussed.

As this book focuses on global IAMs, their geographic scope is unbound and covers the whole human population. On the time dimension, welfare considerations are mostly restricted to the 21st century (though internal time horizons are longer). This cut-off point is somewhat arbitrary since, of course. Well-being does not lose its value at this specific point in time. The reasons are mainly practical: increasing uncertainty and limited computational resources. Given the research question of IAMs, this is a defensible abstraction. Close to all scenarios involve reaching climate neutrality long before 2100, and most costs and consequences of the modeled mitigation choices arise within this century. Nevertheless,

---

[26] A consequent model intercomparison study based on this framework for the first time systematically explored how to meet stringent targets with limited overshoot. Utilizing a discount rate of 2%, it finds, in contrast to the reported costs in earlier studies, the long-term overall costs to be lower with less CDR reliance (*Riahi et al. 2021, 1065*). Riahi et al. acknowledge that the standard IAM framework "by design, favours postponement of mitigation action until later in the century" (*Riahi et al. 2021, 1065*).

[27] Moreover, recent studies concerned with CDR have adopted lower discount rates (e.g. *Fuhrman et al. 2021*).

this domain choice may limit the representation of long-term consequences, for instance, for options such as nuclear energy or some Carbon Dioxide Removal techniques.[28]

The other value judgments in the domain choice concern the focus on human welfare in evaluating climate policies (*Mintz-Woo 2021b, 532*). Ethically, many philosophers agree that nonhuman animals have moral importance. Such a view is especially prevalent in the utilitarian tradition, from which welfare economics emerges. As nonhuman animals experience pain and pleasure, we should consider their well-being in our moral consideration (cf. *Singer 2009 [1990]*; *Regan 2004 [1983]*). Other ethical traditions have recently also argued for the moral recognition of nonhuman animals (cf. *Korsgaard 2018*; *Nussbaum 2022*). However, nonhuman animals find very little consideration in the debate on climate ethics or climate policy in general (*McShane 2016, 2018*). IAMs largely neglect nonhuman interest in determining mitigation pathways, as these interests do not appear on the market and, thus, are not represented in consumption.[29]

One might be inclined to defend integrated modeling in this respect by pointing out that nonhuman interests are taken care of indirectly since mitigation pathways presuppose climate goals. Thus, ethical considerations concerning safe environments are part of mitigation pathways. However, different climate mitigation strategies also clearly affect nonhuman interests in different ways. Take, for example, the extensive reliance on BECCS in many IAM scenarios. BECCS relies on large areas of land used to produce bioenergy. In the AR6 scenarios, around 5% of cropland is used. Such an extensive land demand may lead to rivalry with the habitats of wild animals and thus affect their interests. The anthropocentric scope of IAMs remains thus a value judgment in scenario evidence from mitigation pathways.

As there seems hardly a practical way to include nonhuman interest directly into the IAM framework, such interests must be considered exogenously to the

---

[28] IAMs also neglect these long-term consequences by relying on high discount rates. High discount rates make consequences in the far future appear very small in terms of present values. At a discount rate of 5%, costs of $1.000.000 arising in the year 2300 only count for slightly more than a dollar in present value. Long-term consequences are thus de facto excluded from the calculations even without the temporal cut-off point.

[29] There is little problem awareness about excluding nonhuman animals. Dasgupta, for example, seems to equate non-anthropocentrism with giving nature intrinsic value: "The ethical viewpoint I explore here is self-consciously anthropocentric. Nature has an intrinsic value, but I ignore it because the three books on the economics of climate change I am responding to ignore it" (*Dasgupta 2008, 144*).

models. Side effects on nonhuman animals and long-term risks must be considered when deciding on thresholds for mitigation options such as BECCS and similar means, which involve impacts on the biosphere.

6.8 Handling Uncertainty and the Case of Carbon Dioxide Removal

The final value judgment in IAMs that I discuss goes back to a classic argument concerning values in science: the argument from inductive risk. The inductive risk argument states that scientists must rely on nonepistemic values when deciding whether the evidence is sufficiently strong to support a particular choice in the research process. In such cases, the practical consequences of being wrong must be evaluated, implying that nonepistemic values are widespread in science. In integrated modeling, cases of inductive risk appear in the various choices on data and parameters that must be made when modeling mitigation pathways based on the available evidence. This section focuses on one particular case to explicate the value judgment involved. It also aims to put a new perspective on the much-discussed issues of Carbon Dioxide Removal (CDR).[30] This section will argue that large-scale reliance on CDR poses a case of inductive risk and that ethical analysis of the inductive risk would suggest a cautious approach to modeling CDR. However, past modeling was somewhat optimistic concerning empirical assumptions on CDR and thus involved a particular bias in favor of the present generation.

CDR is an umbrella term for various techniques that "remove CO_2 emissions from the atmosphere" (*Minx et al. 2018, 3*).[31] What unifies different techniques under the label CDR is that they can provide "negative emissions," which promise to compensate for emissions in sectors that are hard to mitigate and to allow shifting mitigation requirements into the further future within the models (*Fuss et al. 2018, 3*). Including CDR in IAMs leads to more flexibility in meeting stringent climate targets, which gives rise to CDR's high economic value in the models. The most prominent types of CDR in IAMs are Afforestation, Bioenergy with

---

[30] This section is a condensed version of the argument provided in Hollnaicher (*2022*).

[31] CDR has been central to recent discussions surrounding IAMs. Cf. for example Fuss et al. (*2014*), Geden (*2015*), Geden (*2015*), K. Anderson and Peters (*2016*), Peters (*2016*), Shue (*2017*), S. Beck and Mahony (*2018a*), Lenzi (*2018*), K. Anderson (*2019*), and Dyke, Watson, and Knorr (*2021*).

Carbon Capture and Storage (BECCS), and Direct Air Capture (DAC). Discussions have mainly concerned BECCS as it has been scaled up most extensively in mitigation pathways.[32]

BECCS has been an established part of integrated modeling for years and has recently featured in most IAM pathways. For example, the scenarios in Rogelj, Popp, et al. (*2018*) rely on negative emissions between 150–1200 $GtCO_2$ across the 21st century. The upper end of this range assumes that humanity can remove the cumulative CO_2 emissions from 1980 to this day out of the atmosphere sometime later this century. However, even comparably moderate rates of 5 $GtCO_2/yr$ would require an industry of the size of today's oil industry, ramped up in just a few decades (*Strefler et al. 2018*). The problem with such large-scale reliance on BECCS in the models is that it is highly uncertain if such a massive scale-up is feasible. BECCS still needs to be developed on a large scale, raising many ecological, social, institutional, and environmental feasibility concerns (*Masson-Delmotte et al. 2018*, Table 5.11). Expert assessment such as Fuss et al., for example, estimates the sustainable potential to be in the range from 0.5 to 5 $GtCO_2/yr$ by 2050 (*Fuss et al. 2018, 14*), much lower than the 5 to 15 $GtCO_2/yr$ assumed in most scenarios runs.

At least some part of this bet on CDR involves empirical uncertainty. Chap. 4 distinguishes three kinds of uncertainty in IAM results. *Epistemic uncertainty* describes uncertainty relating to how the target system functions. *Parametric uncertainty* describes that our model does not perfectly match the target system. *Societal or scenario uncertainty* describes that there is uncertainty about whether the scenarios represent our actual human choices. The first two kinds of uncertainty are empirical and, thus, potentially give rise to inductive risk.

As Richard Rudner (*1953*) originally introduced it (cf. also *Churchman 1948*), inductive risk occurs when scientists need to decide whether to accept a hypothesis based on the available evidence. As evidence for a scientific hypothesis is typically not conclusive, scientists must decide whether the available evidence is sufficient. Rudner argues that the threshold of sufficiency depends on the context. He writes: "How sure we need to be before we accept a hypothesis will depend on how serious a mistake would be" (*Rudner 1953, 2*), and this is, in turn, an ethical question. Heather Douglas showed that similar cases occur in other steps of the

---

[32] I will also focus on BECCS since it has been the most prominent form of CDR (cf. Minx et al. (*2018*) and Fuss et al. (*2018*) for an excellent introduction to CDR). IAM studies involving DAC have seen a recent rise (*Marcucci, Kypreos, and Panos 2017*; *Realmonte et al. 2019*; *Gambhir and Tavoni 2019*), and as DAC phases in even later (after 2050) and scales up rapidly, there are related ethical questions in reliance on DAC (*Fuss et al. 2018, 9*). These arguments may apply to DAC even more if included without explicit limitations.

scientific process, such as in collecting data or choosing a particular methodology (*Douglas 2000*).[33]

The way to analyze cases of inductive risk is by comparing the "foreseeable consequences" of two possible kinds of error: "For any given test, the scientist must find an appropriate balance between two types of error: false positives and false negatives. False positives occur when scientists accept an experimental hypothesis as true, and it is not. False negatives occur when they reject an experimental hypothesis as false and it is not" (*Douglas 2009, 104*). When scientists can mitigate both errors, this is, of course, most desirable, but this is often not possible within a given inquiry. One can only trade off risk between the two kinds of errors (*Douglas 2009, 104*). It is these trade-offs that give risk to value judgments. Values "weigh the significance" of the uncertainty and determine what kind of evidence is sufficient to go forward with a hypothesis or certain interpretation of data (*Douglas 2009, 97*).[34]

Returning to the issue at hand, the parametrization of CDR involves cases of epistemic and parametric uncertainty. In BECCS, for instance, photosynthetic processes capture atmospheric CO_2 in biomass, which is ultimately removed from the atmosphere by being stored underground. BECCS performs exceptionally well in the modeling environment since it produces energy (or other valuable products) and removes CO_2 from the atmosphere. However, the beneficial evaluation of BECCS in the model depends on a range of parameters that must be drawn from the wider scientific evidence. For instance, modelers must decide on the yield rates of crops, which determine biomass production. High yield rates lead directly to higher carbon capture rates of BECCS and further allow the use of more land for BECCS by freeing up areas from food production. Based on technological developments and dispersion, IAMs typically assume high and exponentially rising yield rates for the future. However, such high yield rates have also been questioned by scientists due to ecological and social concerns (cf. *Creutzig 2016*). Moreover, heatwaves and water shortages could considerably

---

[33] This can be viewed to preclude a certain response to the argument already given by Jeffrey (*1956*). He argues that scientists should hedge their hypothesis with the probability of being wrong (cf. *Betz 2013*). However, the outlook for avoiding value-ladenness is small if the inductive risks are manifold and deeply embedded. Rudner anticipated this point and responded that scientists must decide on the hedged hypothesis, which gives rise to value judgments again (cf. *Rudner 1953, 4*; *Douglas 2009, 85*). Individual scientists may use conventions and collective standards of significance to avoid individual, subjective influences (*Levi 1960*; *Wilholt 2009*).

[34] Douglas deemed this influence of values (including nonepistemic values) legitimate since their influence is only "indirect." Values are here not taken to support a hypothesis directly, which would be illegitimate, a case of letting moral conviction determine facts.

impact yields negatively. There is epistemic and parametric uncertainty on how bioenergy will work out in the future compared to their modeled representation.

Inductive risk implies that this uncertainty gives rise to implicit value judgments being embedded in making this parameter choice one way or the other. The inductive risk scheme demands evaluating the foreseeable consequences of error to understand this value judgment. We can merge the background assumptions on BECCS for this analysis in two stylized categories (cf. *Hollnaicher 2022, 4*): *high CDR potential* (due to favorable assumptions) and *low CDR potential* (due to less favorable assumptions). Modelers can be wrong in two ways: either they underestimate CDR, relying on lower potentials than there is in reality, or they overestimate CDR, relying on high potential that does not materialize. Scientists must evaluate these two errors.

The case of *Underestimating CDR* has the consequence that pathways assume higher mitigation burdens in the near future than would be necessary. As they wrongly exclude the option of later emission removal, they assume an unnecessarily steep reduction in the near term. This implies higher costs and burdens than would be economically optimal, which fall primarily on the current generation and arguably on high-emitting countries in the present.

Overestimating CDR, on the other hand, implies that we end up with higher emission levels than compatible with the temperature goals assumed, as we wrongly assumed to be able to remove emissions later. CDR pathways temporarily exceed no-CDR emission pathways by up to 9 $GtCO_2$, one-third of total emissions (*Lenzi et al. 2018*). In this error, the bet on CDR does not materialize; thus, these excessive emissions cannot be removed in the future. Future generations will end up with tremendously steep mitigation requirements or must accept higher warming levels and, consequently, a more dangerous world. Increased impacts will affect predominantly vulnerable parts of the population.

Comparing these two stylized outcomes provides a comparably clear tendency. Underestimating CDR leads to additional burdens and costs for us, who face this choice, whereas overestimating CDR involves shifting risks to the future and creating a more dangerous world for the future poor. Shue (*2017*) describes such betting on CDR as an apparent case of injustice, as the current generation relieves themselves from ambitious mitigation but thereby forces the future poor into dire circumstances: "To keep our own jewellry now we risk forcing others to sell their blankets later" (*Shue 2017, 214*). This is a case of "intergenerational buck-passing" (*Gardiner 2006*), and, as it involves relieving ourselves from duties based on wishful assumptions, it can be seen as what Gardiner (*2011*) calls "moral corruption."

If this ethical analysis is correct, scientists should be relatively conservative when choosing parameters and assumptions for BECCS and comparable techniques (cf. *Hollnaicher 2022, 5–6*). This analysis starkly contrasts how BECCS was represented, at least until recent model adjustments. Creutzig writes that "[c]ommon assumptions chosen in IAMs [...] display only a small sample of the overall assumption space, focusing on the corner of technological and political optimism" (*Creutzig et al. 2015, 8*). The expert assessment of Vaughan and Gough (*2016*) similarly puts seven out of ten assumptions on BECCS within what they call a "danger zone": assumptions with high influence and uncertainty.[35] Other scientists questioned if BECCS will even turn out to reliably provide negative emissions in practice at all once broader impacts and soil degradation are taken into account (*Fajardy and Mac Dowell 2017; Harper et al. 2018; Fajardy et al. 2019; Brack and King 2020*). If modeling results in IAMs depend on such optimistic parameters, the inductive risk analysis implies an especially questionable implicit value judgment that comes with it. Such scenarios put much risk and burden on the future generation to relieve current emitters from high costs. This is a particular and one-sided answer to the question of distributive and intergenerational justice and represents an ethically contentious bias in scenario evidence from IAMs.

To sum up, the example of BECCS showed how instances of empirical uncertainty in technical parameters give rise to value judgments. IAMs rely on many such technical and empirical parameters in modeling climate futures. Weyant writes: "Current IAMs generally make one fairly homogeneous set of simplifying assumptions about risk attitudes, and the implications of alternative assumptions are generally not explored in any depth" (*Weyant 2017, 127*).[36] The case of CDR showed that such parametrization can involve highly contentious value judgments.

---

[35] Further social concerns are described by Fridahl and Lehtveer (*2018*) and Butnar et al. (*2020*). The social and institutional task of managing a sustainable uptake of several GtCO2 yearly, spread across various regions, implicitly assumed by the models, is enormous. In exploiting full CDR potential, the models assume a globally uniform carbon price and robust international coordination—far from the current political reality.

[36] There is much more to be said about the representation of CDR in the models. McLaren and Markusson (*2020*), for instance, analyzes the co-emergence of a policy framing and this kind of modeling framework, a "co-evolution between policy and politics, modelling and science-based technological promises" (*McLaren and Markusson 2020, 395*). Pielke (*2018*) describes how "the models that have analyzed the more ambitious policies have been pushed towards implementing more optimistic assumptions about the range and availability of their mitigation portfolio." This led to the paradox that "despite little progress in international climate policy and increasing emissions, long-term climate stabilization through the lens of IAM [integrated assessment models] appears easier and less expensive" over time.

6.9   Summary

This chapter revealed a range of value judgments in IAMs and feasibility assessments based on them. First, modelers must make value-laden choices concerning indicators and concrete constraints for feasibility. As a salient example, I analyzed the proposed indicators for the economic dimension in Brutschin et al. (*2021*). I showed that mitigation costs and stranded investments involve value judgments, as they tend to favor particular interests and policy strategies. I then turned to agenda-setting and argued that value judgments in this stage slip into later stages of how modeling results come to matter. Value-laden methodologies influenced the neglect of low-temperature targets before Paris 2015 and the kind of pathways that came after.

After these two broader value aspects, the following sections revealed a range of value judgments about how costs are represented in IAMs. *First*, the cost minimization framework is a significant value commitment. It is the main criterion for burden sharing within the models. It represents a value judgment contrasting alternative burden-sharing principles in the justice literature. *Second*, the definition of well-being as consumption is a central value judgment. One alternative approach would be to consider objective elements of well-being, such as whether basic needs are met. *Third*, how well-being is aggregated into welfare is a value judgment. Mostly, SWFs have a utilitarian form, which neglects important ethical aspects due to existing inequality. *Fourth*, discounting is a significant value aspect, and current evidence mainly relies on high discount rates that shift considerable burdens to the future. *Fifth*, choosing a particular domain represents a value judgment within the welfare approach.

Finally, modelers must make many choices concerning parameters, background assumptions, and general model behavior that they have to draw from the broader evidence. Such choices are often underdetermined on empirical grounds and, thus, according to inductive risk arguments, involve value judgments. I have analyzed this concerning the particular case of representations of BECCS. I argued that within the technical and epistemic parameters, modeling of BECCS involves value choices, and, as it stands, IAMs tend to favor present emitters over future bearers of climate change.

"An option was created, not in the real world, but in models that sustain the current policy envelope" (Tavoni/Scololow 2013).

While there might be more value aspects in IAMs, the analyzed aspect represents critical value judgment, which often remains implicit background assumptions for scenario evidence from IAMs. This chapter revealed that scenario evidence from IAMs is deeply value-laden. The next chapter discusses how to deal with such value judgments to retain legitimate and objective advice.

Objective Assessments with Value-laden Models

The last chapter explicated and discussed various value judgments in assessing feasibility with IAMs. This chapter discusses how scientist should deal with these value aspects in their assessments. It proposes three principles that should guide scientist, when they aim for objective and legitimate policy advice with value-laden models: VALUE TRANSPARENCY, VALUE PLURALITY, and DEMOCRATIC ENDORSEMENT. As I shortly discussed in Chap. 4, we should not understand objectivity in the sense of value-freedom and "getting to the things" in modeling feasibility. Nevertheless, objectivity remains a vital goal in scientific assessments that inform policymaking, one that is closely connected to political legitimacy. The three guiding principles I propose in this chapter respond to the problem of legitimacy that arises from value-laden assessments and they, in conjunction, provide a provisional solution to it, or so I will argue.[1]

The first section discusses the meaning of objectivity. It extracts two aspects relevant to achieving legitimacy in scientific policy assessments: objectivity as neutrality achieved on a communal level and objectivity of values. Sects. 7.2 to 7.4 provide three guiding principles: making value judgments explicit premises (Sect. 7.2), increasing the plurality of value positions in IAM evidence (Sect. 7.3), and seeking democratic endorsement of value judgments through venues of deliberation (Sect. 7.4).

---

[1] I lend the term "provisional solution" from *Luckner (2005)*, who uses the term with reference to René Descartes to describe prudential theories as providing guidance in situations where we otherwise lack sufficient normative orientation.

© The Author(s) 2025

S. Hollnaicher, *Assessing Feasibility with Value-laden Models*,
https://doi.org/10.1007/978-3-662-70714-2_7

7.1   Objectivity and Legitimacy in Scientific Assessments

Scientific assessments of feasibility are vital, and to count as scientific, they must aim to be objective. The term *objectivity*, however, can mean very different things. Chap. 5 discussed methods of empirical validation of background assumptions that relied on a sense of "getting to the things." This proved to be conceptually and methodologically confused in respect to using models to assess feasibility. Moreover, the discussions of value judgments in IAMs made clear that objectivity as value-freedom is untenable for guiding assessments with IAM. However, we cannot let go of objectivity altogether as an aim of science and scientific policy advice. This section argues that this is due to the role objectivity plays in ensuring the legitimacy of value-laden scientific assessments and policy advice. This section will discuss certain aspects of objectivity and legitimacy that ground three guiding principles for dealing with values in integrated modeling.

Political legitimacy is the moral permissibility to implement political decisions.[2] An institution has political legitimacy if it is in the moral position to make decisions and enforce them with instruments that would otherwise be objectionable, such as the use of coercion or force (cf. *Buchanan 2002*; *Peter 2023*). Legitimate institutions have political authority. (The two concepts are roughly two sides of a coin.) An institution has authority if we all have moral reasons to comply with its decisions, even in cases where we disagree with them and even if we are right in our disagreement (*Kolodny 2014a, 197*; *Christiano 2008*). The puzzle of democracy is how we can actually come to legitimate decisions. As we all are equals and have the same political standing, no person should permanently be subject to the authority of anybody else. Democracy is "rule over none" (*Kolodny 2014a, 2014b*). In a democracy, all affected people should make decisions of collective interests, each given the same voice (cf. *Fritsch 2019*).[3]

---

[2] This is a normative understanding of legitimacy. There is also a descriptive concept that refers to an institution's *perceived* legitimacy (*Peter 2023*).

[3] Less than half of the world lives in a democratic society. The democratic argument is a normative argument for the proper relationship between the people and their governments, independently of the actual political system in place. It assumes that political power is ultimately only *morally* legitimate in a democratic form. This position is prominent in political philosophy, held either on instrumental (cf. *E. Anderson 2009*) or intrinsic grounds (cf. *D. M. Estlund 2008*; *Viehoff 2017*). Instrumentally, political systems are legitimate if they reliably make good decisions, and democracies are taken to be epistemically superior. Intrinsically, democratic procedures are the only legitimate form of government as they ensure central values such as political equality (cf. *Christiano 2004*; *Viehoff 2014*; *Kolodny 2014a, 2014b*).

However, in real-world democracies, there are people who, in their respective *roles*, have more of a say. We rely on a temporary transfer of power to representatives who make decisions on behalf of us. Such an institution of representation is legitimate, but only if there are appropriate mechanisms of authorization and accountability (cf. *Pitkin 1972*; *Fearon 1999*). We *authorize* policymakers in elections and other forms and hold them *accountable* by various mechanisms, including elections, that are part of a functioning democracy.[4] Scientific experts should have more of a say when it comes to matters in which empirical facts play an important role. Since we rely on scientific knowledge in our collective decisions, the producers of this knowledge might thus be said to have more influence on policy decisions by matter of providing the relevant facts.

Legitimacy is foremost a political concept, but it applies to science, especially scientific policy advice, in a derivative way. To understand this, it is helpful to reconsider objectivity as "value-freedom." Value-free objectivity can be understood as the view that "all values (or all subjective or 'biasing' influences) are banned from the reasoning process" (*Douglas 2004, 459*). Douglas rejects this view as infeasible and, as argued, it is unavailable for assessing feasibility with IAMs. If it were available, however, value freedom would ensure that scientific advice is legitimate. Empirical facts need to be combined with values to make policy decisions. If scientists were to provide value-free facts, no influence on policy decisions beyond their epistemic authority would arise. As scientific knowledge, though, includes value judgments in various ways, the problem of legitimacy arises in science. Scientists lack any special political authority, as they are not accountable to the electorate nor authorized by them in any similar way as political representatives. Scientists therefore might exert an illegitimate influence on political decisions if expert advice is bound up with specific value outlooks. The infiltration of seemingly neutral and objective facts with values risks smuggling in preferences or values in policy decisions.

As the scientific process involves value judgments in various ways, we need a new solution for ensuring that scientific knowledge has no illegitimate influence on policy decisions. For this reason, Holman and Wilholt (*2022*) introduce what they call the "new demarcation problem" regarding values in science. They argue that once we let go of the value-free ideal, something needs to take its place, distinguishing three purposes of the value-free ideal that need to be met: *Veracity*, the orientation of science to truth, *Universality*, the ability to use scientific results

---

[4] It is important to note that both concepts have seen extensive debate and have been expanded beyond a simple election view (cf. *Parkinson 2003*; *Mansbridge 2003*; *Saward 2009*).

independently by all sides, and *Authority*, hindering undue influences of scientist on policymaking *(Holman and Wilholt 2022, 214)*. It is primarily the last two purposes that are directly grounded in the democratic argument presented above. Scientific knowledge needs to be universal so that all sides of a debate can rely on reliable knowledge. Further, an illegitimate influence of scientists must be avoided, as no person should have a greater say in political decisions than any other. In light of these considerations, we need to reassess how to understand the objectivity of scientific advice in the context of integrated modeling. The following sections will go through several aspects of objectivity for this purpose. It will mainly rely on an understanding of objectivity that plays out on the collective level of science.

One crucial aspect of objectivity for this purpose is as a form of policy- or value-neutrality. Douglas describes that in one meaning, something can be called objective if it is "balanced or neutral with respect to a spectrum of values" *(Douglas 2004, 460)*. This meaning of objectivity is often taken as a direct implication of the democratic argument given above: "Value neutrality was intended both to ensure that value judgments were made by democratically accountable bodies and that scientific experts were seen as transpolitical authorities" *(Holman and Wilholt 2022, 215)*. However, what neutrality exactly means in this context is perhaps as contested as the meaning of objectivity.

A place to start in this context is the IPCC mandate. It states: "By endorsing the IPCC reports, governments acknowledge the authority of their scientific content. *The work of the organization is therefore policy-relevant and yet policy-neutral, never policy-prescriptive*" *(IPCC n. d.)*.[5] The mandate plays a central role in the modelers own relation to value and policy questions and has been called

---

[5] Explicit interpretations of the mandate are rare. Havstad and Brown *(2017b)* give an extensive discussion and argue that non-prescriptiveness is the central meaning, with neutrality adding little on its own. Being non-prescriptive means "that they never decisively recommend for or against a policy option—amongst the available courses of action" *(Havstad and Brown 2017b, 6)*. They take the latter parenthesis to have particular importance, as the IPCC is legitimized to exclude "clearly infeasible options." However, this makes their discussion unhelpful here, as we are concerned with what "the available options" actually mean. The examples of clearly infeasible options that the authors give—"e.g., clapping our hands twice in order to cool the atmosphere by 1°C" *(Havstad and Brown 2017b, 6)*—are too fantastical to add substance to their interpretation. The first part of the ideal, policy relevance, is relatively straightforward in its meaning. Policy relevance demands that the scientific knowledge collected and systematized by the IPCC should provide policymakers and the public with information that, in some sense, is important for addressing the various practical challenges arising due to climate change. This shapes the agenda and implies a particular responsibility for the IPCC and the scientists to communicate effectively *(Gundersen 2020, 101)*.

the "defacto governing principle for their [the modeler's] own work for the IPCC and the team of authors with which they collaborated" (*Gundersen 2020, 100*). Low and Schäfer (*2020*) write that "IAM and modeling participants, in response [to the problem of value-laden IAMs], invoke the 'policy relevant but not policy prescriptive' IPCC mandate, and emphasize that the intent of IAM work is 'neutral mapping'" (*Low and Schäfer 2020, 7*). Policy neutrality is not the same as value neutrality, though the two are closely related. Policy neutrality allows for some value commitments. As the work of the IPCC is concerned with the challenges of climate change, it arguably involves the value commitment of being guided by preserving safe planetary conditions and sustainability as a core value of its work. Policy neutrality demands only a neutral mapping of the options that are tackling climate change in some way, but not amongst all options whatsoever (e.g., using enhanced climate change as a new kind of weapon). The meaning of neutrality, as presented in the IPCC mandate, is targeted at reports. A report is objective in this way if it represents all value positions in a balanced and neutral way.

We can apply value-neutrality to policy advice in general. Hugh Lacey has most extensively discussed value neutrality. Lacey (*1999*) defines neutrality as "evenhandedness" in the sense that scientific knowledge is neutral if it is not bound up with any specific value outlook. He writes: "[N]eutrality expresses the value that science does not play moral favorites" (*Lacey 2005, 26*). This means that no scientific theory (in principle) logically entails or precludes any value outlooks. Scientific knowledge should be, in principle, usable in combination with every value position.[6] However, individual studies and scenarios with IAMs will not be usable with every value position, even in principle, as they involve value judgments that cannot be excluded. Moreover, the discussions showed that IAMs involve particular core value commitments, which makes it hard to see how they could overcome these value tendencies completely. For instance, if one completely rejects technologically driven mitigation strategies and holds that climate change should be mitigated by rigid restrictions on our behavior and returning to a nature-based lifestyle, it seems complicated to see how such a person could use the knowledge of IAMs, even in principle.

---

[6] Even as such, neutrality is meant to be compatible with science being *more informative* on some value outlooks than others (in our example, providing more knowledge on the Paris goals than on the attainability of a $+ 5$ °C world). It is only that *in principle* scientific knowledge "can be put at the service of any values, explaining valued phenomena, illuminating the realm of the possible, informing means to ends and the attainability of ends" (*Lacey 1999, 75*).

However, Lacey later revised his ideal of neutrality as he recognized that scientific strategies based on a *particular* value outlook can benefit science overall, for example, in the case of feminist science. In the reformulated ideal of neutrality, not all pieces of scientific knowledge must be neutral *by themselves* (that is, must serve all value outlooks equally), but all value outlooks should be served by some piece of science, *overall* in an evenhanded way (*Lacey 2013, 82*). Individual projects, tools, and scenarios can be informed by particular values (as in the case of feminist science, but also the case of IAM-science) and not only be compatible with neutrality but foster neutrality, as long as they are part of a full exploration of the spectrum of values. Lacey writes that "the primary responsibility of scientists today is to conduct their research within a worldwide body of institutions, with democratic oversight, [...] and to the ideal of inclusiveness and evenhandedness" (*Lacey 2013, 83*).[7] This expanded interpretation of neutrality can be applied to this context. It holds that overall, scientific advice on policy and feasibility must aim for value neutrality in the sense of representing different viable value outlooks evenhandedly.

This revised interpretation of neutrality is achieved on a social level. Objectivity can be understood as a product of the social process of science. The most influential account of objectivity in this vein is brought forward by Longino (*1990*), who argues that the complex interactions between scientists give rise to the reliability and trustworthiness of the outcomes. What makes scientific knowledge authoritative and reliable is not a particular individualistic stance but the workings of a community that is inclusive and responsive to criticism. Longino argues that scientific knowledge is more objective the more it fulfills four criteria: (1) the existence of recognized avenues for criticism, (2) shared standards, (3) community responsiveness to criticism, and (4) equality of intellectual authority (*Longino 1990, 77–78*). Higher degrees of objectivity are directly linked to a diversity of viewpoints present in a scientific discourse. As even empirical

---

[7] Democratic oversight is necessary for the reason Leuschner (*2012*) describes as a problem of circularity in social accounts of objectivity. Objectivity is a product of diverse venues of criticism, but as science needs to be kept together by shared standards and some kind of gatekeeping, these very rules whose criticism is to be attended to and in what way needs to be decided upon. Her pragmatic solution is democratic regulation of science by increasing diversity, etc. This does not mean that there is no restriction on individual projects whatsoever. Douglas emphasizes that this only applies to debates with legitimate value disagreement. "We have good moral reasons for not accepting racist or sexist values, and thus other values should not be balanced against them" (*Douglas 2004, 461*). Most debates in the climate mitigation context, though, can be taken to represent legitimate value disagreements, and thus, value neutrality is a good goal for policy assessments with IAMs.

sciences depend on value-laden background assumptions, such venues of criticism are essential for explicating values and fostering against the influence of subjective judgments influencing scientific results implicitly (*Longino 1990, 72–75*). Objectivity is achieved on the social level of science (cf. *Chang 2012, ch.5*). Research communities with a broader distribution of intellectual authority recognized venues of criticism, and responsive to this criticism from these different sides can be expected to reach greater objectivity in their results (*Longino 1990, 62*). Lack of diversity often leads to unnoticed biases and implicit values embedded in scientific results.

A second noteworthy aspect of objectivity in this area is applying objectivity to values. In metaethics, it is, in fact, a common position to think of moral principles and values as objective (cf. *Ernst 2008*; *A. Miller 2013*). Only a few have defended moral relativism (cf. *Williams 1985*). However, while most theories conceptualize morality to claim universal validity, there is little convergence on a recognized method to detect objective values. There is no overarching method to determine the objective of these values parallel to the scientific method, nor is there an agreed notion of moral expertise. John Dewey (*2008*) provides a rare account of how to determine objective values. For Dewey, public deliberation and the practical testing of values and hypothesis helps to increase their objectivity. Values that survive public scrutiny and prove to solve practical problems are more objective, and thus, we can rely on them. This pragmatist metaethics has been used by Kowarsch and Edenhofer (*2016*) to argue for the possibility of deliberation to achieve more objective policy assessment despite their value dependence. This second sense of objectivity is relevant to this context and is discussed in Sect. 7.4.

This section argues for understanding objectivity foremost in its function to ensure the *political legitimacy* of science, aiming at scientific advice to policymakers and the public. Due to the value dependence of IAMs, objectivity cannot be understood as value-freedom. We need other solutions to the value problem, which ensure that scientific advice is legitimate. Essential aspects of objectivity in this regard are to uphold a commitment to value neutrality, which, though, plays out on the level of the scientific community more broadly. As such, it is compatible with individual projects and communities being based on particular values, as long as all value positions are served and as long as there is sufficient understanding of these value judgments.

This exploration of objectivity for guiding policy assessments with IAMs is, of course, far from conclusive. It, though, shows that there are avenues for increasing the legitimacy of policy assessments with IAMs beyond value-freedom. The

following sections will defend three guiding principles. Each of these principles is a way to increase the legitimacy of expert advice despite its reliance on
unavoidable value judgments. However, as each principle has limitations, neither
principle can guarantee objectivity. Taken together, these principles are meant to
provide a provisional orientation that can be used to address the value problem,
even if no univocal conception of objectivity or theory on values in science is
accepted.

7.2   Principle I: Making Value Judgments Explicit

A common way to increase the legitimacy of scientific assessments is to make
values judgments explicit. If scientists state values as explicit premises of scientific results, the conditional knowledge can be seen as objective as it only
touches on but does not commit to value judgments. This is the first principle of
the provisional solution to the value problem:

1. VALUE TRANSPARENCY: The legitimacy of policy assessments can be
   increased when value judgments in IAMs are made explicit and presented
   as a premise of the scientific outcome.

Transparency is a recognized goal of science, but an emphasis on values is often
missing. At its core, transparency is the condition of an object or process to be
easy to see through. Applied to the context of science, transparency is the quality
of science being done in a way that allows the understanding, reproducibility,
and critique of the structures, assumptions, and histories that underlay a scientific theory or object. Transparency is a matter of degree, as things or processes
are always more or less transparent, and it is often achieved through multiple
complementary forms and venues of transparency (*Elliott 2020*).[8]

Calls to open the "black box of modeling" (*Pfenninger et al. 2018*) are common, and transparency is a stated goal of the IAM community itself (cf. *Skea*

---

[8] Elliott (*2020*) also distinguishes transparency efforts regarding its "agents." While most
transparency discussions have modelers as agents of transparency in mind, there is, of course,
a wider community contributing to the transparency of IAMs. For example, excellent journal-
istic venues, such as Carbon Brief (*Carbon Brief 2023*), explain IAMs and their findings. The
Senses Project by the PIK provides an excellent example of an outreach, explanation, and
visualization project concerning different aspects of mitigation strategies done in cooperation
with communication experts (*PIK 2022*).

et al. 2021; Riahi et al. 2022).[9] Efforts towards greater transparency have been made concerning making the structures and assumptions of IAMs transparent, providing open code, data archives, and robust documentation (*Skea et al. 2021, 6–7; Byers et al. 2022; Riahi et al. 2022*).[10] Whether these efforts are anywhere close to sufficient, given that the models "play such key importance in the development of international mitigation strategy" (*Purvis 2021, 7*), is up for debate (cf. *Robertson 2022*). Many IAMs are not openly accessible and largely intransparent concerning their parametrization and internal structure (*Rosen 2015*). Moreover, the celebrated "open scenarios database" of the AR6 scenarios is difficult to use and lacks input data completely. One can only guess how adequate peer review and thorough interdisciplinary critique of scenarios is possible on this basis. IAM results are still, to a large degree, the result of "black boxes" (cf. *Robertson 2021*).[11]

Of fundamental interest here is *value transparency*. This section argues that transparency concerning implicit value judgments is integral to the legitimacy of policy advice based on IAMs and their assessments. Conventional transparency efforts tend to stay mostly shallow; "deeper forms transparency" would demand making the implicit value judgments involved in the models more transparent (*Bistline, Budolfson, and Francis 2021*). Value transparency is also a direct implication of the self-stated goal of modelers, which demands that users of the scenario should be able to "understand what drives different scenario results" and grasp the "conditionality of results on specific choices in terms of assumptions (e.g., discount rates) and model architecture" (*Riahi et al. 2022*).

Philosophers of science have defended value transparency as an essential goal of policy-relevant science. Martin Carrier (*2021*) argues that it must be the goal of experts operating at the policy interface to "make values visible and subject to explicit judgment" (*Carrier 2021, 9*). Heather Douglas writes that "scientists should strive to make judgments, and the values on which they depend, explicit. [...] Only with such explicitness can the ultimate decisionmakers make their

---

[9] Cf. Bistline, Budolfson, and Francis (*2021*), Boran and Shockley (*2021*), and Robertson (*2021*).

[10] This can be extensive due to the complexity of IAMs. The documentation of the model IMAGE comprises, for example, no less than 360 tight-set pages (*Stehfest 2014*).

[11] However, one needs to be careful in demanding more transparency. Such transparency measures can also become a gatekeeper for more diverse models and scenarios. For instance, the amount of data needed to submit to the IPCC database is extensive, making it difficult for smaller modeling teams to be included in the essential venues for pathways (cf. *Saheb, Kuhnhenn, and Schumacher 08.06.2022*). Difficult trade-offs arise as this can limit the value plurality of IAM scenarios.

decisions based on scientific advice with full awareness and the full burden of responsibility for their office" (*Douglas 2009, 155*). Kevin Elliott argues that "scientist[s] should strive for greater transparency about values as they communicate scientific information so that members of the public can better understand whether or not their values are being served" (*Elliott 2017, 172*; cf. *Elliott and Resnik 2014*).

All three authors rely in some way on the democratic argument for objectivity presented in Sect. 7.1. In this view, the *implicitness* of value judgments threatens the legitimacy of scientific advice. Carrier writes that the "hidden influence" of value judgments is what "is pernicious and makes recommendations illicitly one-sided and misleading. The objective should not be to drive all nonepistemic values out of science-based guidance but to keep facts distinct from values and to make value-judgments explicit" (*Carrier 2021, 9*). Implicit value judgments risk having an illegitimate influence on political decisions. Only when policymakers can see what values are embedded in scientific knowledge can they exercise their office with the "full burden of responsibility" (*Douglas 2009, 155*). If value judgments are explicit, policymakers rely on knowledge in line with their values, legitimately ignore knowledge based on alternative value premises, and commission research with alternative value judgments.

One way to understand how value-explicit scientific advice can be objective is given by what Nagel (*1961*) calls *appraising* and *estimating* value-laden claims. In appraising claims, scientists come to conclusions based on implicit value judgments. For example, when scientists rely on a specific welfare concept and go on to claim that some policy measure reduces welfare, they also appraise this welfare concept. In such a case, scientists make value judgments by accepting a certain normative presupposition. On the other hand, estimates claims touch on value judgment only in an if-then fashion. For instance, if a scientist concludes that given a concept W of welfare, some policy measures would increase welfare W, the scientist does not take an ethical or political stance. They merely assess whether some potential end would be realized by certain options (*Nagel 1961, 495*). Such estimating claims have traditionally been seen as a way to reconcile value-laden science with the value-free ideal of science (cf. *Weber 1904*; *Nagel 1961*; *Betz 2013*). By making ends explicit premises, social scientists can examine social phenomena without committing to the values themselves. Values appear in scientific knowledge but in a conditional way.[12]

---

[12] Cf. Betz (*2013*) argues that scientists should hedge hypotheses with the uncertainty involved to avoid value judgment that would be implied to inductive risk. For criticism of Betz's argument, cf. John (*2015*) and Barrotta (*2018*).

However, while viewing value transparency in this way is promising in theory, there are serious hindrances to achieving such conditional statements in practice (cf. *Alexandrova 2017, 90 ff.*) *First*, value judgments often need to be sufficiently well understood by modelers to present them as separate premises. It is thus no surprise that the IAM community focuses on more tangible objects of transparency, such as opening data, methods, code, and documentation. Scientists also often misperceive value judgments for facts: "More often than not, the ethical value judgments manifested in modeling choices remain implicit; they are subtly implied by judgments that are often perceived by modelers to be of a purely epistemic nature, that is, concerned with only the factual content of the model representation" (*M. Beck and Krueger 2016, 633*; cf. *Carrier 2021, 9*). However, more than making the technical aspect of the models transparent is required, as users of the pathways must understand how the knowledge in question relates to their value perspective. This cannot be achieved by simply reporting parameters or providing technical descriptions. For instance, the AR6 mentions that the discount rate is an influential value-laden parameter. However, it makes no effort to explicate or discuss its value dimensions or related positions in the ethical literature. Ethical literature, in general, is practically ignored by the AR6[13], and leading authors interpret the transparency problem concerning discounting simply as a "[l]ack of documentation" of the parameter (*Skea et al. 2021, 4*).

Second, though, this focus on technical assumptions reflects a fundamental ambiguity and complexity of value judgments. Consider the discount rate, which philosophers often take as the clearest example of a value judgment in the models. There is persistent disagreement between economists and philosophers on this issue. The analysis in the last chapter revealed that the value aspects in discounting are mixed claims, which depend on empirical and normative questions in a closely entangled fashion (cf. *Mintz-Woo 2018a, 2018b*).[14] In such cases, the explication of value judgments will often not allow for clear ethical interpretations, and thus, premising expert advice with value assumptions seems hardly realizable in many instances.

---

[13] The "ethics chapter" in the previous AR5 was a notable exception. This chapter has not seen a successor in the latest reports (and at the time, unfortunately, it was not integrated with the other chapters). Skea et al. (*2021*) neither mention value transparency explicitly beyond citing one relevant paper in passing.

[14] At least among economists, it seems contested whether discounting involves a value question. However, even if accepted, the nature of these value judgments remains often disputed. In a thorough analysis, Mintz-Woo (*2021a, 104–7*) argues that the different parameters in discounting touch on several value questions and mix empirical and value questions.

Finally, IAMs involve simply too many value judgments. Communicating all value judgments as separate premises would make expert advice impossible, complex, and confusing. Even formats favorable to value transparency—think of a small round of experts and very knowledgeable policymakers discussing the implications of mitigation pathways on energy policy (cf. in Berlin2030)— experts need to restrict themselves to communicating the most salient and well-understood value premises of their work. Communicating all value judgments risks overwhelming the audience and thus hindering a clear view of the knowledge.

Therefore, value transparency is only a partial solution to the problem of legitimacy. However, it would also be wrong to be too pessimistic. Achieving value transparency is a social process that involves various venues and agents of transparency (*Elliott 2021*). Value transparency need not be achieved by the modelers alone. It depends on fostering diverse venues of criticism and discussion. Moreover, it is not necessary to make the precise value judgment explicit. Scientists can rely on general ethical principles or narrative value-laden scenarios in communicating the value premises of modeling results. Carrier argues that general principles, such as the precautionary principle, are a sufficient and relatable way of achieving value transparency (*Carrier 2021, 12*).

To underline this suggestion, consider the value judgments of distributing the remaining carbon budget in modeling climate futures, a question that appeared throughout this book. One could model three alternative principles: distributing according to historical responsibility, economic capabilities, and equal-per-capita budgets for the future (cf. *Dooley et al. 2021*). Looking at the principle of historical responsibility, more specific value judgments arise: Which kind of greenhouse gas emissions should exactly count? Should change in land use count as well? Is there a certain cut-off point in time from which to count? Do we understand historic emissions in terms of a fixed, timeless budget, or do we use another method to translate ethical principles into concrete budgets for each nation? Modelers need to specify and provide their methodology, but value transparency can rely on these mid-level principles, which will be subject to interdisciplinary debates. Relying on such general principles also implies room for disagreement and interpretation, thus not wholly relieving scientific results and communication from being value-laden.

There is an influential argument *against* transparency by Stephen John (*2018*) that is worth discussing here. John suggests that transparency and openness are not enhancing but endangering epistemic trust. Laypeople, for the most part, have a false view of how science operates. Their trust in scientific claims, though warranted, is often fragile and based on false premises about the institutions of science (*John 2018, 7*). Making science more transparent would confront this false view and make an open science *less* trustworthy to laypeople. Trust in science, though, is highly valuable to them for relying on true beliefs, even if this trust was based on a wrong institutional view. Therefore, John argues that scientists should refrain from striving for transparency in communicating scientific findings.

While not targeted at values in science, it can be easily extended along those lines. Laypeople often have a false view of science as a body of facts that can be neatly kept apart from value questions. If laypeople were confronted with the value-ladenness of science, they might easily find scientific institutions to be fraud with political influences and thus (wrongly) perceive them as untrustworthy. Communication of scientific results is thus better done without bringing implicit value judgments to the open. This would be an argument against the desirability of value transparency.[15]

However, John's conclusions seem overblown. It is an open question whether the strategy of non-transparency is empirically the most promising to enhance trust.[16] For example, it is unclear why scientists must accept the false folk philosophy of science as a given. Experts can try also communicating how science works and, by such explanation, hedge against the loss of trust John raises. John points to the lack of time and resources "to wait for a better world" (*John 2018, 7*). Given the lengthy reports of the IPCC, this seems hardly convincing here, and John's argument neglects the various venues, agents, and layers of transparency (*Elliott 2020; Lenzi 2019*).

Most importantly, though, the democratic argument for objectivity rests on reasons of legitimacy, which cannot be resolved on instrumental grounds alone.

---

[15] Modelers have also brought forward such arguments. Keppo et al. (*2021*) states that "documentation is not always as helpful for non-experts as one would hope, since the implications of specific assumptions only become clear when one understands the model well. Similarly, making code and data publicly available is valuable, and teams are increasingly doing this, but few people know how to run and critique a model of this kind" (*Keppo et al. 2021, 14*). This breaches scientific integrity, shielding highly policy-relevant models and scenarios from thorough evaluation (*Robertson 2022, 18*).

[16] Non-disclosure comes with its risks. Take, as a recent example, the scientists' assertions that masks are not as effective as protection against the transmission of the virus early in the COVID-19 pandemic. The collective intention behind such communication was to safeguard the limited supply for the clinics' people who needed masks the most. Scientists wanted to prevent a "mask run" and thus instrumentally adapted their messaging. As the message changed, trust in science was hampered by this earlier deceit. Similarly, insisting on a value-free view of mitigation pathways to foster (rightly needed) political action risks precisely the confusion that John warns against.

Hidden value judgments in policy advice would remain a democratic problem, regardless of whether they would cause distrust if they were uncovered. John's argument does not address the problem of legitimacy that implicit value judgments create.

Viewing value transparency as a way to completely reconcile science with the value-free ideal is unattainable, given the complexity and entangled nature of value judgments in the models. However, value transparency remains of vital importance for increasing legitimacy in assessment. Modelers and other agents of transparency should explicate the value premises of different IAM scenarios and investigate their influence. The last chapter revealed a series of ethical assumptions that can guide making general value principles in scenario evidence explicit.

7.3   Principle II: Modeling a Diverse Array of Values

Value transparency must be accommodated by a second principle that demands increasing the *value plurality* in policy assessments. As policymakers need to be able to make good decisions based on their values and in recognition of the available alternatives, the second guiding principle demands a plurality of values in policy assessment for realizing legitimate and objective policy assessments:

2. VALUE PLURALITY: The legitimacy of policy assessments is increased when all viable value positions are represented in scientific assessment on feasibility overall in an inclusive and evenhanded way.

The principle VALUE PLURALITY is based on the intuition that the real problem concerning the legitimacy of IAM assessments is not based on the existence of value judgments at all but that the value judgments currently represented in IAMs are one-sided. According to this principle, value judgments are unproblematic as long as sufficient value plurality exists across models, scenarios, and parameter choices. Recall the revised conception of neutrality by Lacey (*2013*), which argued that value commitments in individual projects are compatible with neutrality *if* all values are served by some piece of science. This must be done *overall* in an evenhanded way (*Lacey 2013, 82*).

Not only is VALUE PLURALITY central to achieving neutrality in policy assessments, but it seems to be a vital instrument for VALUE TRANSPARENCY as well. Enhanced levels of value transparency demand that we see what relative influence a particular value judgment has and what the implication would be if we chose an alternative value position. The dependence goes both ways. Only if we know what value judgments arise in our assessments can we increase their plurality.

Carrier (*2021*) argues for a plurality of values in expert advice. Experts should strive to present "alternative value-laden policy packages, which combine facts, scientific accounts and nonepistemic premises" (*Carrier 2021, 12*) to policymakers. "The expert ambition should [...] be to enable politicians to make good fact- and value-based choices" (*Carrier 2021, 13*). Central to this ambition, in light of the deep entrenchment of values in science, is to provide an "array" of "alternative value-laden policy packages" (*Carrier 2021, 12*). The direct implication of this principle for IAM research is to increase value diversity in IAM pathways. IAMs need to provide evidence of the complete space for social choice. This task would involve increasing scenario plurality, modeling equally speculative technologies, institutional changes, or behavioral changes, and aiming to create knowledge for diverse value positions. Modelers should embrace plurality concerning critical value dimensions since there is no way to keep value choices out of scenario design.

Plurality can resolve the legitimacy problem of value-laden science. When scientific knowledge exists for different value perspectives, policymakers can rely on scientific reports in line with their values and contextualize scientific knowledge based on different value premises. The illegitimate influence of scientists only occurs if the scientific advice is one-sided and biased (and implicit). Combined with VALUE TRANSPARENCY, this principle thus forms the basis for achieving legitimate policy advice. Recall the social account of objectivity, which links scientific objectivity to the diversity of viewpoints in a scientific discourse (*Longino 1990*). The lack of diversity often leads to unnoticed biases and implicit values embedded in scientific results. Policy assessment can be viewed as more objective when more value positions are represented.

However, as widely accepted as pluralist positions are, they have some limitations as a strategy for increasing legitimacy in policy assessments. *First*, integrated modeling is quite resource-intense, and thus, one needs to make hard decisions concerning what scenarios to model. VALUE PLURALITY has been criticized for being overly demanding and practically "unworkable" (*Havstad and Brown 2017a, 114*), as the authors write in response to the framework of Edenhofer and Kowarsch (*2015*). They write that the permutations necessary would become nearly endless, even considering only a few value dimensions.[17] Achieving value plurality thus ends up being "an impossible fiction" (*Havstad and Brown 2017a, 115*).

This would overplay the problem, however. Relying on general principles and value positions limits the value judgments needing modeling, and one can start with the most influential value questions. Moreover, integrated modeling already produces many scenarios. Including more value-explicit scenario definitions is achievable. Further, the principle above limits the demand to model only "viable" value positions. "Viable" means two things here: Value positions are viable if they are logically consistent and have inner cohesion. The narrative structure of scenarios helps to achieve such coherence. Value positions are further viable if they represent at least somewhat "acceptable" positions. Scientists are justified in dismissing some value positions, for instance, judgments that would pose severe violations of human rights or a clear violation of fundamental democratic values such as equality. There is, for instance, no need to represent explicitly neo-colonial futures. Feasibility judgments can dismiss "unacceptable" value judgments. These responses to the resource problem of VALUE PLURALITY, though, imply that there remain value-laden and contested decisions on what to model (cf. *Leuschner 2012*). The IAM community must find ways to determine what values require representation, which will involve value-laden choices.

A *second* limitation arises because IAMs have structural limits regarding accommodating *all* viable value perspectives. The last chapter revealed some serious value commitments of IAMs that seem hard to overcome. For instance, reliance on a quantified approach to social welfare provides a way to make difficult trade-offs and comparisons. However, it will always ignore certain value aspects that are difficult to quantify. IAMs cannot provide evidence on the whole array of plausible value positions.[18] This implies adding other methods of policy

---

[17] Havstad and Brown (*2017a*) throw the baby out with the bathwater, dismissing the well-developed PEM of Kowarsch and Edenhofer (*2016*). They stay vague on a solution and the meaning of "stakeholder participation to begin in earnest" (*Havstad and Brown 2017a, 119*). The paper fizzles out as they criticize Edenhofer and Kowarsch (*2015*) for not properly including deliberative engagement, although this is at the core of the PEM model (cf. *Kowarsch 2016*). The authors, overly diligently, argue that there are more outcomes than the three temperature outcomes Edenhofer and Kowarsch (*2015*) present in their sketch (something that we can safely assume to be known by Edenhofer and Kowarsch).

[18] However, value plurality needs to catch up even within IAM science. For instance, "overall, most scenarios are around the SSP2 socio-economic assumptions" (*Riahi et al. 2022, 309*) in the AR6, even though SSP2 only represents a small subsample of the plausible, value-laden projections of the future.

assessments in key reports and databases, even if they lack the quantitative depth of IAMs.

To conclude, VALUE PLURALITY is central to achieving greater objectivity of scientific assessment based on IAMs. The legitimacy of policy advice is most severely corrupted if the scenario evidence represents only a small part of the value space. Plurality should be a guiding principle of IAM science, but it has limitations, as all principles do. Achieving plurality concerning value judgments is resource-intense and demands value-laden choices to reduce the complexity of the value choices.

7.4 Principle III: Deliberating Values and Pathways

Making value judgments explicit and subject to plurality increases the legitimacy and objectivity of scientific advice. However, both principles have their limitations. Detecting concrete value judgments is often complex and contested, limiting the possibility of presenting them as separate premises for scientific results. Moreover, resource constraints in modeling and the reception of scientific advice limit the extent to which plurality can be realized. The third principle thus emphasizes relying on democratic endorsement for legitimizing value-laden policy assessments:

3. **DEMOCRATIC ENDORSEMENT**: The political legitimacy of assessments with IAMs can be increased when value judgments are endorsed in open and inclusive venues of public deliberation.

This section argues that democratic endorsement can legitimize value judgments in policy assessments but that the form of endorsement must be understood deliberatively. Most recently, Lenzi and Kowarsch (*2021*) defended the value of deliberation in assessing feasibility within the climate context (cf. also *Edenhofer and Kowarsch 2015*; *Kowarsch and Edenhofer 2016*; *Dryzek and Pickering 2019*; *Lenzi 2019*). They point out the extensive value that such specific forms of public engagement can bring to policy assessment in the climate context and rely on hands-on experience of deliberative forums (cf. *Garard, Koch, and Kowarsch 2018*). Such discussions on democratic endorsement take place against the backdrop of new forms of knowledge production (cf. *Jasanoff 2004*) and the general discussions on legitimacy in deliberative democracy (cf. *Habermas 2019 [1992]*; *Parkinson 2003*; *Dryzek 2010*).

There is a form of "democratic" endorsement that we need to contextualize up front. One way to view value judgments in science is to think of them as *social* values and to assume that they are, ultimately, empirical judgments about what value people *in fact* value. If this were true, the problem of scientists would not be a normative problem but an epistemic problem to detect these social values. This neglects the real normative dimension of the value problem. Moreover, the methods scientists use for detecting values are often questionable. Economists tend to think that values are observable social variables that can be derived from people's behavior in free markets. For instance, as quoted above already, Nordhaus argued that we should parametrize the welfare function to align with "the outcome of market" since then economists need "not make any case for the social desirability of the distribution of incomes over space or time of existing conditions, any more than a marine biologist makes a moral judgment on the equity of the eating habits of whales or jellyfish" (*Nordhaus 2013, 1082*). This is misguided, as the last chapter argued in various instances, as the outcomes of the market do not represent social values adequately, nor can such an approach answer the normative question of what values we should use in our policy assessment at all. Another questionable method is the reliance on surveys, which are known to be sensitive to framing issues and political ideologies (cf. *Lenzi and Kowarsch 2021*).

One needs to view this approach as a particular *normative* answer to the value problem. Kevin Elliott argues, for instance, that one aspect of responsible handling of value influences in scientific reasoning is to make them "representative of our major social and ethical priorities" (*Elliott 2017, 10*). He writes, "[w]hen clear, widely recognized ethical principles are available, they should be used to guide the values that influence science. When ethical principles are less settled, science should be influenced as much as possible by values that represent broad societal priorities" (*Elliott 2017, 14–15*). Relying on democratically endorsed values is defended in the "aims approach," proposed by Kristen Intemann. She argues that "social, ethical, and political value judgments are legitimate in climate modeling decisions insofar as they promote democratically endorsed epistemological and social aims of the research" (*Intemann 2015, 219*).

A particularly salient example where this strategy legitimizes value judgments is in "commissioned research." In cases of commissioned research, the values and aims of a particular research project are directly provided by the authoritative body. Carrier (*2021*) notes that in such cases

"[N]ormative commitments are compatible with the value-free ideal if they are adopted by commission. Scientists are commissioned by policymakers to explore ways to achieve certain social goals. In such schemes, the values at hand are set from outside of science and scientists are authorized to make the pertinent value judgments" (*Carrier 2021, 12*).

If we could use this model for policy assessment with IAMs, the problem of legitimacy might be solvable. When authoritative sources provide value judgments within scientific assessment, their influence is no longer illegitimate. The providers of the values and the authoritative body for making the decisions are then aligned, and thus, the democratic problem described above does not arise in any substantive form.

However, such an understanding of democratic endorsement as a simple external provision value judgments is, however, unavailable for many value judgments in assessing feasibility with IAMs. *First*, there is no democratically endorsed value base for most value judgments arising in modeling futures, nor will such value judgments be derivable from the aims of IAM research. This approach would seriously underestimates the value diversity in society, especially considering the global scope of IAMs (cf. *Kowarsch 2016*; *Kowarsch, Flachsland, et al. 2017*). There is reasonable disagreement about most value judgments appearing in IAMs.[19] Social values or priorities are not even undisputed in cases with democratically endorsed policy options, such as the commitment to the Paris Goals or the German decision to exit nuclear energy. Taking democratically underlying value judgments as fixed input for modeling climate futures could be seen as scientists preventing deliberating on these policy options.

Second, even where we find a seeming consensus of social values, these values could still be ethically mistaken. Too often, social values have turned out to be unjust or biased in favor of dominant social groups. Taken as a general solution to the value problem, this strategy would limit the role of science to an executive role, where social values simply need to be taken as given and neglect the critical role of science. Expert advice to policymakers needs to keep a certain degree of independence and explore knowledge and policy options that conflict with the current social aspirations, as Carrier (*2021, 13*) writes.

Understanding democratic endorsement, however, as a deliberative process involves a change of perspective, which can encompass the two worries above.[20] Viewing social values as a given input to policy assessment with IAMs would

---

[19] Also, relying on democratically endorsed values foregoes that only about half of the world is living in democratically organized societies. *Politically* endorsed values by other regimes cannot resolve the value problem outlined above.

[20] Intemann hints at such a deliberative understanding when developing her aims. She writes that the endorsement should be subject to "democratic mechanisms that secure the representative participation of stakeholders likely to be affected by the research" (*Intemann 2015*,

mean regarding democracy primarily in an aggregative fashion (*Young 2000*). Highly simplified, it thinks of the democratic process (as above) only in terms of a fair procedure of aggregating individual preferences into a collective decision. Value positions, then, might be "taken" from the democratic procedures. This view is limited, though, as it considers people's priorities fixed and accommodates only a relatively thin understanding of rationality (*Young 2000, 21*). The deliberative model, in contrast, takes the democratic process primarily as an ongoing public exchange of reasons. Individual and social values are, in this view, open to change.[21] Concrete decision-making might still involve procedures of aggregation. However, primarily, it is a process of collective reasoning that involves all perspectives and is aimed at reaching a common outcome that has the chance of greater public support.

There is an extensive debate on how such deliberation can and should be understood, especially in a large-scale society (cf. *Parkinson 2003*; *M. B. Brown 2018*). An often-mentioned condition for the legitimacy of such processes and their outcome is that deliberative venues are inclusive, open to all relevant perspectives, and conducive to fostering a deliberative attitude (*Garard, Koch, and Kowarsch 2018*). Concrete deliberative forums of public engagement, for instance, are mini-forums with randomly chosen participants engaging on a topic over several sessions with expert inputs (cf. *Kowarsch et al. 2016*; *Lenzi 2018*), deliberative stakeholder workshops, or deliberative polls (cf. *Fishkin 1991, 1997*). Note that such deliberative venues do not necessarily debate values directly but deliberate on policy measures, pathways, and goals concerning a concrete topic (cf. *Blum 2022*). Thus, value judgments are discussed within their application context (*Edenhofer and Kowarsch 2015, 57*).

This allows for deliberation on value judgments deeply entrenched in pathways and mixes empirical and normative aspects. Moreover, deliberation can change social values. Where controversy cannot be resolved, it can provide modelers with different relevant value outlooks as input to further modeling exercises. It also increases understanding of the inherent normativity in mitigation pathways, a process Edenhofer and Kowarsch (*2015*) call "mutual learning."

Relying on deliberative public engagement fosters the legitimacy of value judgments in two ways: *first*, they can be seen as backed up by bodies with a

---

219) and that such processes involve "interactive feedback loops" (*Intemann 2015, 228*) and multiple venues of criticism.

[21] Deliberative exercises have been shown to be influential in reaching shared decisions despite diverse starting preferences. Such deliberative bodies are, for instance, citizen councils. These bodies exist of randomly drawn citizens who deliberate on particular issues in a safe setting to reach a shared recommendation for the government on this policy question.

certain democratic authority.[22] As Lenzi and Kowarsch (*2021*) writes, "inclusive deliberation processes can increase the democratic legitimacy of climate assessments and climate governance" (*Lenzi and Kowarsch 2021, 24*). They help avoid scientific assessments being technocratic. As the last chapters have shown, many policy assessments are based on the ingrained expert judgment that risks foreclosing public choices concerning the future. Public participation can help to detect and address such a narrowing of the feasibility space.

Second, scientific knowledge that survives criticism from diverse perspectives can be seen as more objective, according to the social understanding of objectivity. Lenzi and Kowarsch (*2021, 26*) note that deliberation improves the epistemic quality in two ways. Deliberation can foster value transparency and adequacy. Modelers can be more confident that results do not involve substantive hidden values or are biased if they are extensively discussed with stakeholders with diverse value perspectives. Moreover, deliberation allows public opinion to change along with the demands posed by policy assessment on climate mitigation. Deliberation has the potential to not only detect value and feasibility judgments of the public but to transform them in light of scientific assessments (*Lenzi and Kowarsch 2021, 25*). Note that this provides a starkly different view of engaging on feasibility to the expert-based framework presented by Brutschin et al. (*2021*), discussed in Chap. 5.

This third principle proposes that democratic endorsement in a deliberative form can increase the legitimacy and objectivity of policy assessment. Value judgments should not be seen as observable social values but as something on which there can be open and inclusive deliberation. If done in the right way, this will increase the quality and legitimacy of feasibility assessments.

7.5   Summary

This chapter proposed three guiding principles for legitimizing value judgments in policy assessment with integrated models. These principles are *(1)* to make value judgments explicit and to communicate them as separate premises of the results, *(2)* to provide scenario evidence on a vast plurality of viable value outlooks, and

---

[22] How much democratic authority such bodies have is controversial. Their authority is rooted in their representativeness and epistemic contribution, not in forms of accountability or actual participation of all (*M. B. Brown 2006*). At least in such context, though, public participation can increase legitimacy, as well-made forums will have some democratic authority on value questions in contrast to scientific experts.

(3) to deliberate on value-laden mitigation pathways and policy measures and rely on such deliberation as a form of democratic endorsement of value premises.

The democratic argument provides the basis for these principles for objective policy assessments. As policy assessments depend in many ways on value judgments, and scientific experts lack the political authority to decide value judgments, assessments pose a problem of political legitimacy. VALUE TRANSPARENCY relieves value judgments in assessments from being illegitimate because it makes the resulting knowledge conditional on the value premises and thus not itself value-laden. This strategy is limited as realizing value transparency this way is challenging to achieve in light of contested and mixed claims. VALUE PLURALITY provides legitimacy, ensuring that all viable value positions are represented. This is important as a contribution to value transparency and a way to aim for value neutrality of assessments. DEMOCRATIC ENDORSEMENT provides a third way to legitimize assessment through public engagement with citizens and stakeholders. This promises to increase the objectivity of assessments by disclosing hidden value judgments in assessment and, potentially, aligning assessments with social values and transforming them through informed and open deliberation.

These three principles do not provide a complete framework or worked-out theory. However, they provide avenues for improving policy assessment with IAMs concerning the value problem in assessing feasibility. We might think of them as a *provisional solution* to the value problem in policy assessments, in light of there being no univocal solutions to handling values in expert advice in general, nor being there a reliable strategy that ensures the legitimacy of assessment. The following chapter will briefly discuss the implications of these principles in light of the value questions and judgments in IAMs.

Applying the Principles: Biases and Ethically-explicit Scenarios

8

This final chapter takes stock. The previous chapters have explicated a range of value judgments appearing in IAMs. The last chapter argued that integrated modeling must strive to make value judgments explicit, subject to greater diversity, and engage with them in deliberative forums to retain the legitimacy of expert advice on feasibility with IAMs. This chapter asks where we stand in relation to these principles. It analyzes current scenario evidence from IAMs in terms of three biases (Sect. 8.1). It discusses how this can contribute to perpetuating injustices under the mask of seemingly neutral and objective assessments with IAMs (Sect. 8.2). It discusses a series of implications of the three guiding principles, emphasizing the need to model value-explicit scenarios with IAMs (Sect. 8.3). Finally, it reflects on the metaphor of modelers as mapmakers and concludes that the findings of this book suggest a particular version of this image (Sect. 8.4).

8.1 Three Biases in IAM Evidence

The analysis of Chap. 6 revealed a series of value questions arising in IAMs and in the methodology of assessing feasibility based on scenario evidence. Moreover, the taxonomy showed that influential ethical premises often remain intransparent and one-sided. This section argues that we can think of these implicit value influences in terms of three biases in current scenario evidence from IAMs, as such evidence is used in assessing the feasibility of climate goals. If this analysis is correct, modelers must address these value issues, and users of the pathways should be mindful of them when interpreting what the models tell us.

© The Author(s) 2025
S. Hollnaicher, *Assessing Feasibility with Value-laden Models*,
https://doi.org/10.1007/978-3-662-70714-2_8

A *bias* is a systematic distortion in scientific results that unduly favors particular interests or value positions over others (cf. *Wilholt 2009*). We primarily understand the term "bias" these days with cognitive biases in mind, which are systematic psychological distortions in perceiving and interpreting the world. Biases in science are related to this phenomenon but describe a systematic distortion in our scientific methodologies and practices. Scientific biases can be the result of various influencing factors. For instance, the prevalence of mainly male subjects in medical studies might result in biased risk assessments in drug prescription, which neglect risks that apply primarily to female patients. Such scientific results can implicitly favor male interests if they are relied upon in policymaking, medical practice, or drug approvals. Other cases of scientific bias go back to economic influences in research processes (cf. *Resnik 2000*).

The notion of bias I use is normative, as biases are *illegitimate* value tendencies in results. For the matter of this book, I will characterize a bias as an instance where the three principles for legitimate policy assessment discussed in Chap. 7 are insufficiently realized. Scientific results are biased in this sense if the results favor a particular value outlook or interest over others when this favoring is not due to epistemic reasons or due to an explicit value premise or a well-founded democratically endorsed value influence. Cases of such biases can go back to methodological choices, parametrizations in the models, or background assumptions in the scenario evidence, which favor particular value positions over others. If such choices make the results of the models one-sided and implicitly so, we need to pay special attention to them in light of the principles for legitimate scientific advice.

Presentism

We might name the first bias in IAM evidence *presentism*. Presentism is the implicit favoring of the current generation's interests over the interests of future generations. Presentism in IAMs occurs due to high discount rates and overly optimistic parameters. Most scenario evidence from IAMs used for assessing feasibility relies on high social discount rates of around 5—6 % (*IPCC 2014a, 2022; Masson-Delmotte et al. 2018*). Such high rates shift mitigation burdens into the future, making near-term requirements in the pathways less drastic than they would be if lower rates were applied (cf. Sect. 6.6). The AR6 acknowledges that "[l]ower discount rates <4% (than used in IAMs) may lead to more near-term emissions reductions" (*Riahi et al. 2022, 305*).

As IAMs typically include Carbon Dioxide Removal as a mitigation option, high discount rates also fuel the reliance on negative emissions (*Emmerling et al. 2019*). If relied upon as a conventional mitigation option, carbon dioxide removal allows a significant transfer of burdens to the future. The concrete technological representation of CDR aggravates the presentist tendencies of high discount rates. As discussed in Sect. 6.8, concerning BECCS, IAMs involve parameters and assumptions that are comparably optimistic regarding how well these future technologies can provide negative emissions. I argued that the inductive risk of such parameter choices is disproportionately borne by vulnerable people in the future, who would bear the consequences of aggravated climate change if the technologies fail to deliver. Therefore, parameter choices involve an implicit value judgment favoring the present emitters. For instance, assumptions on BECCS assume relatively high yield rates and low estimations of side effects in biomass production. Reviewers of the SR1.5, thus, lamented that IAM pathways were "strongly biased in favor of BECCS" (*Hansson et al. 2021, 5*). Large-scale reliance on negative emissions involves a highly unequal risk profile, in which current efforts are evaded by shifting risk to future generation (*Shue 2017*; *Lenzi 2018*) (cf. Sect. 6.8).

Further, risk profiles between social groups and generations are often unequal when other significant parameters are only represented by a median value. More precautionary parameter choices would often protect vulnerable groups. If this is so, they might add to the presentist bias. For instance, the climate sensitivity is often represented by the median value of around 3 K to interpret mitigation pathways in light of the warming that would result. The latest calculations still involve considerable uncertainty, however, and state, for instance, a "up to an 18% chance of being above 4.5 K (7% in the Baseline calculation)" (*Sherwood et al. 2020, 94*). This could alter the knowledge on how it is feasible to stay within the Paris Goals tremendously. Other parameter choices in IAMs similarly culminate around median values. In the AR6, for instance, "most scenarios are around the [middle scenario] SSP2 socio-economic assumptions" (*Riahi et al. 2022, 309*). Often, risks of error in such choices are born disproportionally by future generations akin to the analysis given in Sect. 6.8.

Existing scenario evidence, thus, arguably involves a *presentist bias*, which implicitly favors the current generation by shifting burdens in mitigation pathways into the future.

Privileging Welfare in the Global North
A second bias is a *privilege for welfare in the Global North* within IAMs. The reliance on efficiency, the disregard of justice-related considerations, and the reliance on technological and growth-based strategies lead to an implicit favoring of countries from the Global North.

From the standpoint of justice, different lines of reasoning suggest that high-emitting countries from the past have a greater responsibility for bearing the burdens of mitigating climate change. Greater obligation can be seen as a consequence of their historic responsibility, as they have used up more of the carbon budget than countries from the Global South (and benefited from doing so). Moreover, they tend to have greater resources, partly due to the exploitation of countries from the Global South and nature and less need to address urgent social issues such as poverty or food crises. Both responsibility and ability support greater responsibilities on parts of the Global North (cf. *Shue 1999*; *Caney 2010, 2018*; *Dooley et al. 2021*). These considerations are also central to the UNFCCC Charta, which declares to "protect the climate system [...] on the basis of equity and in accordance with their common but differentiated responsibilities and respective capabilities" (*UNFCCC 1992*, Article 3.1). In contrast, results from PB-IAMs often show considerable shifts of burdens to the Global South due to the supposed global and intertemporal cost-effectiveness. This seemingly neutral efficiency criterion gives way to an implicit privilege of welfare in the Global North.

One way to express this tendency is in terms of the remaining carbon budget for different regions. IAM studies generally do not model an explicit distribution of the remaining carbon budget but distribute the remaining emissions where they provide the most welfare in terms of marginal utility gains by consumption. The implicit neglect of justice considerations makes IAM scenario evidence, for the most part, fall on comparably unjust distributions. For instance, the IAM pathways in the AR5 exhibit remaining emissions below 50 $GtCO_2$ for Latin America, below 150 $GtCO_2$ for the Middle East and Africa, while remaining emissions in the OECD region are around 400 $GtCO_2$ in AR5 pathways (*IPCC 2014b, 435*), concluding that the "Middle East and Africa (MAF) region and especially Latin America (LAM) have the largest mitigation potential" (*IPCC 2014b, 434*). In the AR6 scenarios, "emissions are typically almost equally reduced across the regions" (*Riahi et al. 2021, 335*), implying that in 2050, emissions in the Global North are still equal to the emissions in Africa and the Middle East and considerably higher than in Latin America where high amounts of CDR are projected and lead to earlier Net-Zero dates than in other regions.

The point here is that these distributions allow the Global North to continue with relatively high emissions and smoothen their transition on the cost of legroom for the Global South. CDR adds considerably to this implicit privilege of the Global North. AR6 scenarios remove between 200 and 600 GtCO2 via BECCS, using around 5 % of total global cropland (*Riahi et al. 2022, 347*). As biomass potential is often higher in the Global South, especially in Latin America, least-cost pathways shift significant amounts of mitigation burdens to these countries (*Riahi et al. 2022, 324*). Mitigation pathways, thus, implicitly rely on land in the Global South to account for a shrinking carbon budget, for which the inaction of high-emitting countries is disproportionately responsible. By neglecting justice considerations, major injustices become ingrained in scenario evidence under the mask of "neutral" cost optimization. Applying explicit fairness criteria to CDR deployment in the models, for instance, shows that two to three times higher CDR efforts would be necessary under these assumptions in OECD countries in 2050 (*Strefler et al. 2021, 8*). The neglect of such considerations, thus, shapes mitigation strategies in the model tremendously.

This bias can also be witnessed in many background assumptions concerning energy demand and lifestyle patterns. For instance, a considerable energy privilege for the Global North is assumed within mitigation pathways from IAMs. In average IAM scenarios, energy consumption as late as 2100 in "OECD countries and the rest of Europe is 2–3 times more than the average energy consumed in the Global South" (*Hickel and Slamersak 2022, e629*). Scenarios in which energy consumption equalizes until midcentury are missing from the scenario space of IAMs. The existing scenario evidence from IAMs involves a persistent pattern of greater consumption in the Global North, which gets inscribed into feasibility assessments if not sufficient other visions of the future are added. Existing scenario evidence implicitly favors the welfare of already privileged countries.

Favoring Entrenched Interests
A third bias in scenario evidence and the reliance on it in assessing feasibility is the *favoring of entrenched interests*. IAMs and the methodologies of assessing feasibility tend to favor the status quo. As described in Chap. 5, model evaluation often uses empirical data from the past to adjust and parametrize the models. Such model adjustment to improve the representation of the target system promises to make models more "realistic" and not too far off from precedents. However, as the past involves substantive climate inaction at the cost of the future and more vulnerable people, these injustices risk being inscribed into constraints of feasibility in the future. Favoring the status quo is implicitly ingrained when feasibility assessment relies on "appeals to the past," and when existing structures and distribution are taken uncritically as a baseline. Brutschin et al. (*2021*), for instance, uses stranded fossil investment as a direct indicator for feasibility concerns. Feasibility assessment based on such indicators favors the interests of fossil industries over other interests, as discussed above.

Moreover, IAMs are structurally built to represent gradual change, limiting their ability to represent a "revolutionary overhaul of the system" that some say is needed at this stage of climate governance (*K. Anderson 2019, 348*). Lastly, favoring the entrenched interests is visible in the ubiquitous reliance on economic growth in IAM scenarios despite a growing public discussion on alternatives. Keyßer and Lenzen (*2021*) describes that "[n]one of the 222 scenarios in the IPCC SR1.5 and none of the shared socioeconomic pathways projects a declining GDP trajectory" (*Keyßer and Lenzen 2021, 2*). Such tendencies to favor the status quo come from modelers' perceived need to keep a certain proximity to the existing discourse (*van Beek et al. 2022, 200*). However, in doing so, mitigation pathways from IAMs are implicitly biased in favor of entrenched interests.

These three biases play out in different dimensions of justice. The first bias concerns questions of intertemporal justice, the second questions of global justice, and the third questions of social justice. I argued that evidence from IAMs involves substantive bias in all three. In this context, I defined a *bias* as a systematic favoring of certain value positions or interests in which dependence on value premises is not made explicit and not subject to plurality within the scenario space on feasibility. This analysis neither covers all value tendencies in the existing scenarios for climate governance nor can it speak to all modeling studies. If this analysis is correct as a general assessment of mitigation pathways, however, it heightens the urgency of addressing the lack of plurality, value transparency and public deliberation in IAMs. Moreover, the adequacy of relying on the scenario evidence from IAMs to assess feasibility is, by this analysis, put in serious doubt. The following section will argue that these biases risk perpetuating injustices under the guise of feasibility if they remain unaddressed.

8.2   Perpetuating Injustice under the Guise of Feasibility

The last section explicated three general biases in current IAMs. Policy assessments should not uncritically rely on scenario evidence from mitigation pathways, especially when such evidence is applied to the feasibility question. This section describes how these biases risk perpetuating existing injustice under the guise of feasibility if used uncritically.[1]

There are two senses of how injustice is perpetuated under the mask of feasibility. In the *first sense*, value assumptions are built into the models, often under

---

[1] This formulation is inspired by Hickel and Slamersak (*2022*).

the guise of what is deemed to be *politically feasible*. Take, for instance, the fact that there were so few low-emission pathways before 2015 as more ambitious climate targets were seen as infeasible by many at the time (cf. Sect. 6.2). Here, implicit beliefs on the political feasibility of more ambitious near-term mitigation and technological scale-up led modelers to produce evidence almost exclusively on less ambitious climate targets. Only in recent years have modelers focused on more disruptive mitigation pathways.

Alternatively, take the value judgment using Negishi weights (see Sect. 6.5), limiting the global wealth redistribution within welfare-optimizing IAMs. The internal tendency of welfare-optimizing IAMs is to align income levels globally. Modelers view such tendencies to be politically infeasible and therefore use Negishi weights to freeze the income distribution between regions. This is done to achieve a "more realistic" model behavior. However, a significant value assumption enters IAMs under the guise of this judgment of realism. In both instances, value judgment gets embedded in scenario evidence from IAMs under what modelers deem politically feasible. These value judgments arguably land on corners of the available value space that are rather unjust, as they benefit groups privileged by the current distribution of income and burdens. Injustices are slipped in under the guise of feasibility.

One problem here is, of course, that modelers are arguably not in a good *epistemic position* to make these kinds of judgments. Judgments on the political feasibility of different options beyond the immediate future are generally highly uncertain. Consider, for instance, the recent build-up of LNG terminals in northern Germany. How many observers would have judged it as feasible for Germany to build LNG terminals within just a few months? Given the energy crisis after Russia broke war on Ukraine, this quickly turned out to be feasible. Judgments of feasibility are highly context-dependent and often involve an unstated understanding of background conditions. More often than not, we hide what we are willing to do under cover of what is infeasible. Such feasibility claims are often made with little empirical support, and given the discussions of Chap. 4, it seems unclear that good evidence can be provided at all. Modelers are arguably no experts in questions of political feasibility (cf. *Lenzi and Kowarsch 2021*).

The second problem is that these judgments take a position on *contested value questions*. Take, for example, the sustained long-term energy inequality in IAM scenarios. A political realist might hold that powerful nations will always act in their best interest and, thus, not give up on their high energy demand (cf. *Posner and Weisbach 2010*). Modeling realistic pathways, thus, would demand sustained high energy consumption in the Global North, which implies a wide disparity in energy consumption with the Global South. This feasibility judgment can be empirically challenged. Why should the Global South give up on their interest in growth, especially given a shifting power balance in the world? Feasibility judgments in either direction are a rather bold claim. However, more importantly, assuming energy disparity and taking the opposite position of greater equalization, both represent very different value outlooks. It is best to treat such judgments as value judgments.

The *second sense* of perpetuating injustice under the guise of feasibility concerns the framing and presentation of the results. Chap. 4 argued that modelers rely on the framing of assessing feasibility to demarcate their work as scientific and objective. Modelers use such conceptual framing to make clear that they are not taking any stance on the desirability of different strategies. In doing so, however, modelers miss the critical dimension of value in the pathways, which gets implicitly engrained in their judgments on the feasibility of climate goals and mitigation strategies. I argued that modelers must explicate, diversify, and engage with value judgments. If not, modelers risk perpetuating injustice under the guise of feasibility in this second sense by wrongly presenting scenario evidence as value-free and neutral assessments of feasibility despite the described biases. When feasibility assessments are brought forward by scientific experts in authoritative reports, such as the IPCC, this could contribute to continuing unjust power structures of the world under the guise of neutral scientific feasibility facts.

These two senses of perpetuating injustice under the guise of feasibility are practically relevant. Integrated modeling is influential in providing visions of the future for policymakers and the public discourse. Integrated modeling is central to the scientific discourse on solutions and mitigation strategies. IAMs have "a place in the sun" within the current climate governance (cf. Sect. 6.6). Social scientists describe this as a "performative" nature of IAM pathways as IAMs shape the policy discourse (cf. S. Beck and Mahony 2017; Haikola, Hansson, and Fridahl 2019), pointing out that IAMs have "world-making power [...] by providing new, political powerful visions of actionable futures" (cf. S. Beck and Mahony 2018b). Thus, the IAM community has a special responsibility to avoid biases. Moreover, framing integrated modeling under the heading of assessing feasibility carries heightened requirements. IAM science risks being complicit in maintaining injustices by bringing forward encompassing knowledge on mitigation that is biased and accommodates the expected unwillingness of the governments of the Global North. The modeling community must urgently counter this tendency not to fail the most vulnerable.

Assessing feasibility based on current evidence risks perpetuating injustices under the guise of feasibility facts. Value judgments become ingrained in scientific evidence on the back of implicit judgments of feasibility, and value judgments risk being smuggled into policymaking by presenting value-laden evidence as neutral and objective feasibility assessments.

8.3   For Explicit Ethics in IAM Scenarios

Existing evidence from IAMs involves biases and thus risks perpetuating injustice under the guise of feasibility. Modelers should emphasize making value judgments explicit, subject to a greater variety, and interdisciplinary and public debate. This book has, though, yet to address how greater transparency and plurality can be achieved concerning the value dimension of IAMs. While I cannot give a comprehensive framework for doing so, this section draws some implications from the principles.[2]

The *first* implication is the need to *model scenarios explicitly with ethical questions* in mind. This book laid out a range of value questions appearing in IAMs. To make the value premises of IAMs explicit and understand their implication, modelers should vary these assumptions, targeting value aspects directly in scenarios. The discussions showed that, in many instances, the fact that modelers ignored value questions and relied on seeming shortcuts to circumvent value judgments had the consequence of incorporating *implicit* value judgments in IAM evidence. The implicit influence of value judgments leads to the problems above. Providing ethically explicit scenarios can help address this situation.

Such value scenarios must engage directly with the various ethical questions arising in modeling mitigation pathways. By preserving a non-committal attitude, modelers can investigate such scenarios without taking a stance on the underlying value question. For instance, IAM pathways could model different principles for a fair distribution of the remaining carbon budget. Ganti et al. (*2023*) define ethical variations of distributing the remaining carbon budget. Dooley et al. (*2021*) provide a comprehensive discussion of competing principles. Such principles can be operationalized in IAM scenarios and used to highlight the mitigation easement countries of the Global North enjoy in efficiency-based pathways compared to a more just effort sharing.

---

[2] Systematic approaches to improve integrated models in this regard have been defended by other authors (cf. *Edenhofer and Kowarsch 2015*; *Kowarsch 2016*; *Workman et al. 2020*, *2021*).

One example of such modeling can be found in Strefler et al. (*2021*), which models the implications of different fairness principles on deploying Carbon Dioxide Removal in IAMs. The authors compare least-cost pathways with scenarios based on historical responsibility and capability.[3] The study shows that if countries of the Global North take their fair share (under different interpretations of fairness), they must develop two to three times more CDR than in globally cost-efficient pathways. Such results are highly policy-relevant and put the existing scenario evidence in perspective. To make value-explicit scenarios work, one can rely on mid-level objects, such as different policy strategies or operationalized principles, as a proxy for ethical outlooks. Value transparency in such cases would be significantly improved if they integrated ethical discussion of these principles and ethical literature in general. This will require transdisciplinary work with climate ethicists.

A *second* implication is that modelers should *contain the influence* that *implicit* value-laden parameters have if they cannot be made sufficiently transparent. Currently, a few parameters are in the unfortunate corner of greatly influencing scenario results *and* having limited value transparency. However, often, there are ways to limit the influence a value-laden parameter has, for instance, by defining scenarios in more detail. For example, discount rates are a major driver of high CDR deployment in mitigation pathways. Responding to widespread criticism, modelers proposed limiting the influence of the discount rate on CDR deployment by making the amount of CDR an explicit scenario parameter (cf. *Rogelj et al. 2019*; *Riahi et al. 2021*). So far as this is applied, it turns contentious policy implications into an explicit premise of the scenario. While discounting remains an implicit value parameter, its influence is contained as it no longer determines the dependence on future build-up of CDR. Such a strategy could also be used for other influential value judgments. I argued, for instance, to rely on explicit global distributions of mitigation burdens to constrain the influence of the efficiency framework.

A *third* implication is that there is a critical role for philosophers and ethicists (cf. *Kowarsch and Edenhofer 2016*; *Lenzi and Kowarsch 2021*). Ethicists are not moral experts in the sense of being providers of authoritative value judgments for policy (cf. *Mintz-Woo 2021a*). However, they are normative experts who can detect and provide different arguments for investigating ethical questions. As IAM needs to achieve greater transparency and plurality concerning value judgments

---

[3] Another example of value-explicit scenarios is, for example, the project "Just Transitions to Net-Zero Carbon Emissions for All (JustTrans4ALL)" (*Zimm, Schinko, and Pachauri 2022*; *Pachauri et al. 2022*).

in IAM pathways, ethicists can play a critical and constructive role in modeling the climate future. Lenzi and Kowarsch (*2021*) distinguish three roles: in a *diagnostic* role, ethicists identify normative claims (*Lenzi and Kowarsch 2021, 3*). In a *justificatory* role, ethicists help find "the most compelling normative reasons" for value judgments (*Lenzi and Kowarsch 2021, 3*). Both these roles are important, but the authors highlight a third, *collaborative* role for ethicists in which they co-determine feasibility sets for policies. Since assessments rely heavily on ethical assumptions, philosophical ideas could provide new scenarios and connect value scenarios more to normative literature (*Lenzi and Kowarsch 2021, 6*). Collaborative work has already diversified the scenario space of IAMs, for instance, by bringing in questions of distributive justice to modeling (cf. *Budolfson et al. 2017, 2021*). Ethicists need to engage with value questions in integrated modeling in all three roles.

A *fourth* implication addresses malpractice in some modeling and policy assessments. Often, greater objectivity is sought by averaging value judgments. However, taking the average of two ethical principles makes the resulting judgment no more objective. This is already observed by Max Weber when he writes: "But this has nothing whatsoever to do with scientific 'objectivity.' Scientifically the 'middle course' is not truer even by a hair's breadth, than the most extreme party ideals of the right or left" (*Weber, Shils, and Finch 1949, 57*). Scientific contributions would circumvent value questions by proposing a seeming compromise or settlement of different possible value perspectives on a question. For instance, the Climate Action Tracker uses a "weighting scheme" to model different distributions of the carbon budget in order "to make sure that all equity viewpoints (categories) are considered equally" (*Climate Action Tracker 2019b*). This is certainly a possible normative method, but it is no more objective as a solution to the value problem than settling for any single principle but another value judgment. There is no way to shortcut normative aspects in modeling the future (cf. *Dooley et al. 2021*).[4]

A *fifth* implication is that the IAM community should foster the plurality of their research community. Social accounts of science view objectivity as achieved only on a community level. Longino (*1990*) points out how objectivity depends on the presence and recognition of diverse viewpoints in a scientific discourse. This is important since the presence of diverse perspectives helps to bring implicit value judgments in background assumptions into sight and makes them subject

---

[4] Note that this is one answer to the question of "moral uncertainty" in policy assessments. For the general discussion on moral uncertainty, cf. Lockhart (*2000*), Harman (*2015*), and MacAskill (*2020*). For a more specific discussion, cf. Mintz-Woo (*2018a*).

to debate. Integrated modeling, however, is mostly conducted by institutions in countries of the Global North. As described in Chap. 4, more than half of the scenarios go back to six main models, all located in Europe and the USA. (Of course, though, not all researchers are from these countries.) Few research hubs are influential in coordinating the community, leading important modeling projects, and serving as coordinating authors of the IPCC reports. Given the ethical biases above, this congregation of influence might be taken to have negative epistemic consequences. The IAM discourse arguably represents a "very Western vision of the world" *(Saheb, Kuhnhenn, and Schumacher 08.06.2022)*. If we follow the social accounts of scientific objectivity, one measure to address this is fostering diverse criticism venues and heterogeneity of viewpoints within the research community.[5]

A *sixth* and final implication from this discussion is that IAMs cannot cover the whole value spectrum. Value plurality in IAMs faces pragmatic and structural limitations, as the taxonomy has shown. Realizing value plurality implies the need to embrace other methods for assessing feasibility. While IAMs are versatile and can extend their value spectrum, they still inhibit a particular perspective that is hard to overcome. The crucial point is that even if IAMs have specific value commitments that they cannot overcome, this is compatible with the neutrality of scientific advice in general, as long as other value-laden methods are equally represented in science and central reports. Currently, IAMs have a favored role, for instance, being disproportionally represented in IPCC reports. Alternative methods for envisioning the future are often more readily seen as dependent on a particular value perspective. Given my analysis, integrated modeling similarly involves a particular stance. Value plurality, thus, demands including other value-laden scientific perspectives, for instance, by including more non-IAM pathways in IPCC reports and the scenario database and by contextualizing IAM results in light of implicit value judgments.[6] The IPCC's primary reliance on mitigation pathways from IAMs in assessing the 1.5 °C goal is problematic in this light (cf. *Hansson et al. 2021*). Feasibility knowledge should not be conditional on one particular value perspective.

---

[5] The underlying question is who gets to discuss feasibility. As feasibility is an influential political concept, this framing heightens the stakes of public involvement and diversity of the research community (cf. *Leuschner 2012*; *Eigi 2019*).

[6] There is an ongoing debate in the philosophy of science on whether bringing in some value perspectives is detrimental (*Biddle and Leuschner 2015*). Plurality has some constraints, and while I cannot discuss them in more detail, my principle only demands to include all viable and democratic perspectives. Compare, for instance, discussions on a "moderate" form of scientific pluralism Cartwright (*2006*).

These implications directly result from applying the three guiding principles presented in Chap. 7 to the state of IAMs. While staying short of a framework for realizing greater transparency and plurality, they provide insights into how IAM science can improve.

8.4   Being Mapmakers, Reconsidered

This final section of the book returns to the widespread metaphor that was mentioned in Chap. 4. An established picture modelers use to convey the objectivity of their enterprise is that they are "mapmakers" for future paths of society, with policymakers being the navigators that ultimately determine the course to take. This image carries an aspiration. It conveys that modelers aim to provide relevant, neutral, and objective policy advice in the "largely unknown territory of climate policy" (*Edenhofer and Minx 2014, 37*). This section will not challenge this cartographic image but will develop it further. It invites us to reconsider *what kind of maps* we should think of in viewing modelers as mapmakers.

Edenhofer and Minx introduce the map metaphor for integrated modeling in response to criticism of the political influence in the approval of the Summary for Policymakers of the AR5. They describe that the report is a "living map" developed in a "social learning process between scientists (mapmakers) and policy-makers (navigators), to be used to traverse the largely unknown territory of climate policy" (*Edenhofer and Minx 2014, 37*). They write that it is crucial to keep in mind the clear distinction between the "legitimate roles of scientists as mapmakers and policy-makers as navigators" (*Edenhofer and Minx 2014, 38*; cf. *Edenhofer and Minx 2014*; *Kowarsch 2016*). Edenhofer and Kowarsch (*2015*) explain that this framework responds to the value-dependence of science while existing models of the science-policy-interface fail to answer the "philosophical challenges regarding implied value judgments and the objectivity issue in assessments" (*Edenhofer and Kowarsch 2015, 57*).

However, the metaphor of mapmakers has gained a life of its own. Modelers emphasize the map's objective and neutral qualities when relying on this metaphor to delineate their work from the value-laden sphere of policymaking. In this sense, mapmaking anticipates what is relevant for policymakers but does not make value judgments themselves. It provides a neutral and objective map and thus stays aloft of value questions. In Low and Schafer's interview study, for instance, the mapmaking metaphor is used as an interpretation of the "'policy relevant but not policy prescriptive' mission" and stated in conjunction with neutrality as "neutral mapping" (*Low and Schäfer 2020, 4*). Haikola, Hansson, and

Fridahl (*2019*) describe a purist version of the mapmaking metaphor in which "modellers should concern themselves only with drawing the maps, as accurately as possible, for the navigating policymakers to use" (*Haikola, Hansson, and Fridahl 2019, 10*). For modelers, the map image provides a way to distinguish their scientific work from the political sphere. Low and Schafer observe that IAM modelers then "emphasized the advisory, 'map-making' function of their work" (*Low and Schäfer 2020, 4*), using the map metaphor in what is called "boundary work," the social negotiation of a boundary between legitimate and illegitimate forms of scientific practices and advice (*Gieryn 1983; M. Beck and Krueger 2016; Haikola, Hansson, and Fridahl 2019*). The map metaphor draws the boundary that critics attempt to dissolve by pointing out the political nature of modeling.

It is easy to see how maps can convey the image of objectivity. The objectivity of the map goes back to its potential to represent the geographic surface accurately and consistently. Maps are objective in the sense of that they "get to the things." If this were our guide for integrated modeling, the long-term objective of modelers as mapmakers might envisage an ever more accurate and versatile map comparable to the map provided by Google. Users of Google Maps can zoom in on any part of the globe and navigate the streets of Lilongwe as fluently as the streets of Bielefeld. This map is highly accurate in representing different aspects of reality. Concerning a science of feasibility, such a map would accurately represent the totality of the relevant causalities for mitigating climate change. Users could rely on models understood this way to orient themselves about different aspects of the needed transformation.

The latest developments in integrated modeling might be understood in this aspirational sense. Modelers have developed an open scenario database that allows systematic access to all output data for the AR6 scenarios (*Byers et al. 2022*), and outreach projects provide graphically appealing browsing of different scenarios (cf. *PIK 2022*). Moreover, the latest developments in assessing feasibility aim to provide a systematic and scientific way to comprehensively assess the various dimensions of feasibility (cf. *Brutschin et al. 2021*). The underlying sense of objectivity is closely related to value freedom and getting to an independent reality, which, as this book has argued at length, is unattainable concerning feasibility constraints and evaluations.

Critics of IAMs see the map metaphor rather negatively, as they view it as providing a too narrow image of what integrated modeling does, framed in an image representing reality. Beck and Oomen, for instance, prefer to speak of modelers as "corridor makers" that *narrow* the space of possibilities to a corridor in what they see as a "performative practice" (*S. Beck and Oomen 2021, 170*). van

Beek et al. (*2022*) describe that the mapmaking function is in a dilemma with the widespread anticipation of "policy no-go's," which gives rise to the risk "that modellers exclude transformative pathways that contain politically challenging but potentially crucial low-carbon strategies" (*van Beek et al. 2022, 200*). Critics thus reject the map metaphor as it masks the performative and value-laden nature of integrated modeling.

However, once we think about maps more closely, it becomes evident that there are all kinds of maps. Topographic maps, showing the elevation and other visible surface elements, are only one map type among many. There are *political maps*, highlighting the borders of nation-states and other political units, *military maps*, showing defensive lines and troop placements, *thematic maps*, portraying geographic areas with a particular subject in mind, *linguistic maps*, showing the distribution of languages, and, to use the example from Kitcher in his reflection on science in general, the transit maps such as the *London Underground Map*, which shows the station and lines of the Tube (*Kitcher 2001, 56*). These differences reflect that maps are built for a purpose and that this purpose implies depicting different elements of reality. Mapmakers apply different abstractions and idealizations to the landscape depending on their purpose. This does not preclude maps from being accurate or objective.[7] A transit map can accurately represent the train lines, even if the geographic distances and directions are wrong. A political map shows wrong elements compared to the geographic surface, drawing lines into the landscape and painting areas in different pictures. However, it is accurate in relation to its standard.

If we recognize the vast variety of maps, we need not abandon the map metaphor, but we can embrace it. What makes a map objective is that the standards applied in the map are clear, explicit, and consistently applied. The map users must be aware of these standards, which Kitcher calls the "reading conventions" of a map. If the standards with which the map is drawn are transparent, the different kinds of maps can be seen as accurate and reliable. Take, for instance,

---

[7] Kitcher, in developing his account of objective science, points out that wholly different maps can be described as accurate: "If practical success in navigating is to serve as our test of accuracy, then the map of the London Underground must count as accurate—for it figures in the successful activity of tens of thousands of people each day" (*Kitcher 2001, 57*). Even though the map abstracts from many features, for instance, the geographic distance between two stations, it accurately represents reality in light of its purpose. These reflections on mapmaking illustrate that there "is no unique correct way for a map of the globe, or of some smaller region, to draw boundaries" (*Kitcher 2001, 58*), and similarly, there is no reason to believe in science to provide context-independent truths. However, even if there is no universal true theory, we have instruments to describe some theories as more accurate or true, the same way we can ascribe maps to be accurate despite vast differences.

the category of cartograms, which display the land area of a country in relation to an explicit thematic standard. For instance, there are maps showing the size of the country in relation to its historic emissions. In such maps, the USA, Europe, and China balloon to more than their usual size, while Africa and South America shrink dramatically (cf. *World Mapper 2019*). Such maps are common and convey a different image of mapmaking. While such a map would look only vaguely related to the earth's geography, it can be objective as a map, as long as the standards for drawing the map are explicit. Such maps can be highly useful if the information mapped is relevant to us.

If we view scenarios from IAMs in light of this variety of maps, it becomes evident that the maps modelers provide involve an explicit value standard. This book argued that a wholly empirical grounding of background assumptions and parameters in IAMs is unattainable. The inherent value-laden standard of IAM's mapping must be transparent, plural, and subject to deliberation. Moreover, as integrated modeling provides one general *kind* of mapping, we must embrace other methods to assess the feasibility of climate goals and mitigation pathways. However, scenarios that address ethical questions relating to different mitigation futures can provide informative and objective maps of these issues and help us better understand the realities of feasibility.

Conclusions

9

I started by claiming that this book concerns the *value problem in assessing feasibility*. The question raised by this problem is how value-laden models can produce assessments of feasibility. The short version of my answer is the following: IAMs are in a good position to give scientific advice on the feasibility question relating to climate goals. Assessing feasibility demands an integrated perspective and dynamic pathways, which IAMs provide in a particular way. However, IAMs need to engage more directly with their internal normativity to provide legitimate and trustworthy assessments of the feasibility question. This means that modelers should aim to make values more *transparent*, include a *greater plurality* of value perspectives, and *deliberate* upon value aspects with the public. These three strategies help to safeguard modelers from what would be an illegitimate influence on policymaking. While this cannot free modeling from value judgments, such value commitments are compatible with neutral advice as long as other value-laden methods are equally represented. Currently, though, scenario evidence from welfare-optimizing IAMs dominate assessments of feasibility and the models involve a range of implicit and one-sided normative assumptions. In many instances, modeling studies shortcut or mask value assumptions on a false sense of neutrality and objectivity. In practice, this leads to current evidence from IAMs involving three biases under the veil of feasibility. Feasibility assessments need to address this urgently.

The three parts of the book contributed in different ways to this outcome. *Part I* argued that feasibility should be understood as the attainability of an outcome given the resources and processes at our disposal. Feasibility is about there being a viable path. I argued that feasibility is a thick concept as it involves contextual assumptions concerning what kind of means, side effects, and uncertainties are

S. Hollnaicher, *Assessing Feasibility with Value-laden Models*,
https://doi.org/10.1007/978-3-662-70714-2_9

morally acceptable. The conceptual debate showed that assessing feasibility in relation to complex collective goals requires us to take an overarching perspective, which is attentive to all constraints simultaneously and allows for dynamic pathways for realizing an outcome to count towards feasibility. Finally, feasibility is normatively consequential as judgments of infeasibility can be used to rule out specific options or goals in our practical deliberation.

Part II argued that PB-IAMs are in a suitable position to fulfill the presuppositions of feasibility claims. IAMs integrate all relevant dimensions and produce transformation pathways, allowing for complex dynamics to achieve the goals. IAMs have a central role in advising policymakers on climate change. With the adoption of the Paris Goals and in light of more demand for knowledge on solutions, modelers recently developed systematic ways for assessing the feasibility of climate goals and mitigation pathways. I argued that scenarios from IAMs can serve as evidence for the feasibility of climate goals but that this relation of evidence depends on the validity of certain background assumptions. Modelers attempt to evaluate these assumptions empirically, but the concrete methods used involve conceptual and methodological shortcomings. Most severely, appealing to the past in determining background assumptions and constraints for assessing feasibility risks excusing agents due to their past unwillingness.

Part III contributed to the outcome by providing a taxonomy of value judgments in IAMs and by discussing how modelers should deal with values in policy-relevant assessments. It explicated and discussed the various ways IAMs depend on normative assumptions. The taxonomy showed prevailing value judgments in the current scenarios and hinted at alternatives for each value aspect that could be implemented in the modeling framework. The value-laden nature of IAMs raises a problem of legitimacy for policy advisors. Modelers risk undue influence on policy decisions when policymakers rely on their assessments without being aware of implicit value assumptions. To hedge against illegitimate influence, modelers must pursue efforts to make normative assumptions explicit, provide evidence on the whole array of viable value perspectives, and deliberate upon them with policymakers and the public. The last chapter described the urgency of this demand. The currently available evidence from IAMs involves three biases: scenario evidence from IAMs tends to favor current generations, favor the welfare in the Global North, and favor entrenched interests. If this were not addressed, modelers would risk perpetuating existing injustices under the veil of what is presented as neutral and objective advice on the feasibility issue. A step to guard against this is to model scenarios with specific ethical questions in mind, contributing to the transparency and plurality of modeling studies.

This is the analysis this book provides on assessing feasibility with IAMs and on how to respond to the internal normativity in order to achieve more objective and legitimate assessments. This book is certainly not a complete investigation of values in integrated modeling and assessing feasibility, nor do I think my arguments give concluding answers to these debates in any way. The scope of my analysis is limited in a range of ways: *First*, this book discussed PB-IAMs deliberately under the framing of "assessing feasibility." The aim was to connect the scientific practice of integrated modeling with the philosophical discussions on this contested concept. However, there are different ways to make sense of integrated modeling. Other uses of the models are not directly addressed in this book, and the reader must consider how the analysis applies to these uses. Moreover, this book included interdisciplinary work on how to make sense of IAMs to a smaller degree than it deserves. It is a contested question how to understand the models and their results. This book is primarily a philosophical contribution to an interdisciplinary field.

Second, this book simplifies discussions on the models in various ways. There are many different PB-IAMs, which are highly complex and heterogeneous in their representation of different aspects of mitigation. To be able to discuss value judgments and make implicit tendencies visible at all, my book neglected much complexity. This implies that concrete value judgments will turn out differently for different models and be more appropriate to some of them than to others. Moreover, IAMs present a moving target, and as modelers respond to the critical debate on the models, some concrete details of this book will invariably have a short half-life.

Third, the substantive ethical discussions in this book engage with the models from a particular standpoint. This book did not provide an all-encompassing landscape of value judgments in IAMs. For instance, much more could be said concerning value judgments in technological assumptions and how they would fare within my analysis. Moreover, much more must be said concerning the different viable value alternatives on a particular question and their possible justifications. However, this would get out of hand quickly. I spent more time on a few instances where I considered the discrepancies between implicit normative assumptions in IAMs and widely held beliefs in climate ethics most substantive. Such selection, discussion, and provision of alternatives to value judgments invariably involve a value perspective on its own. Bringing in this value perspective contributes to understanding the models' normative dimension. However, it should not be mistaken for a detached or neutral analysis of the models. It is but one small contribution to the larger, interdisciplinary debate needed on values in IAMs.

Let me also discuss an objection one might have about how I discussed the models. One could object that speaking of the value-ladenness of the models misperceives the kind of knowledge the models provide. As laid out in Chap. 4, IAMs provide scenario knowledge. Their conclusions, thus, take the form of if-then-judgments. In this case, it could be seen as misleading to speak of value-laden models and assessments. The knowledge from IAMs is always conditional. If the value judgments are seen as a part of the if-clause, the conditional judgments that are the product of IAMs may not properly count as value-laden. This is the objection. If we take this objection to imply that the value problem is not substantive but concerns only a problem of interpretation of modeling results, it is, however, a too simple response to the value problem.

First, besides explicit scenario assumptions, a range of background assumptions go into modeling futures, many of which are value-laden. These background assumptions support drawing conclusions from scenario evidence on feasibility judgments in the real world. It is essential to evaluate these background assumptions, and we must do so with respect to value questions. *Second*, even scenario premises on technological and social aspects often lack transparency and plurality regarding *values*. If my arguments are convincing, without greater transparency regarding the value dimension, simply reporting scenario parameters and documenting the models are insufficient for the policy-relevant assessments IAMs produce. *Third*, IAMs provide compelling visions of the future. Therefore, what knowledge is produced matters. If we only have knowledge on some solution strategies, even if conditional, feasibility assessments still have a value problem.

However, in a way, this objection is right. Models produce if-then-assessments, and users of the pathways need to be able to understand what assumptions the models depend upon. Given their policy relevance and central role in assessing feasibility in authoritative reports, as the IPCC reports, *increasing understanding of the conditionality of modeling results* is paramount. This book argues for paying greater attention to normative aspects of the antecedents of such results.

This book provided ideas and insights into the three strands of literature I mentioned in the introduction. It brought a different angle to the *interdisciplinary discourse on IAMs*, by analyzing the models with particular attention to conceptual aspects of feasibility and under the lens of normative judgments. This book critically sheds light on recent attempts to frame modeling as an empirically grounded feasibility assessment. The methodologies of such assessment neglect central aspects of the concept, including concerns going back to its consequential normative role. A second contribution is the taxonomy and discussions of concrete value judgments in PB-IAMs. This taxonomy also suggested alternative value outlooks to be considered in IAMs and showed that feasibility assessment cannot solely rely on IAM scenarios. As IAMs depend on a particular value perspective that seems hard to overcome, the aims of transparency and plurality demand to contextualize their results further.

This book provided a thick conception of feasibility to the literature on *feasibility and non-ideal ethics*. It showed that, at least in this particular field, the proliferation of feasibility judgments by the sciences is complex and value-laden. Thick feasibility questions the conceptual literature's widespread commitment to *Descriptive Feasibility*. Moreover, this book could inspire contributions on concrete normative aspects in integrated modeling. While there have been some contributions on CDR, it would be highly valuable if non-ideal climate ethics would engage with more value aspects in modeling climate futures with PB-IAMs.

To the *philosophy of science* discourse, this book extensively investigated value judgments in a particular class of models. It argued for a provisional solution to the value problem, which might also apply to other forms of scientific policy assessments. It discusses the goals of value transparency, plurality, and democratic engagement and provides a thorough application of these principles by critically analyzing value judgments in a particular field.

These contributions imply interesting avenues for future research. The most concrete avenue of research that follows from these discussions is an investigation of concrete ethical aspects of integrated modeling. The taxonomy is one small step to a research agenda that engages with the value aspects of the models more thoroughly. Such contributions would profit from cooperation between researchers from climate ethics and modelers to tackle concrete value questions arising when providing pathways for the future.

A further research question concerns the value dimensions of the concept of feasibility. This book provided an instance of the proliferation of feasibility facts, which showed the deep entanglement with value judgments. I suggested ways to distinguish between legitimate and illegitimate value influence in feasibility claims and ways to combat illegitimate influence from values in policy assessments. Much more could be said concerning the implications that can be drawn from the philosophy of science debate to the conceptual discourse. The debate on conceptual engineering illustrates that conceptual work must not be limited to conceptual analysis. Can and should philosophers thus align the conceptual discourse on feasibility closer to a thorough understanding of the scientific practices that provide these kinds of facts (cf. *McTernan 2019*)? How is this compatible with the critical role of philosophy concerning scientific practices?

Reading this book as a case study can inspire questions in the philosophy of science as well. An open question remains how value transparency and plurality can be realized in light of the difficulties of detecting, understanding, and communicating ethical assumptions, as these difficulties appear even in the relatively straightforward case of modeling futures. I consider deliberative practices and interdisciplinary work necessary, but how this should be understood more concretely must be examined (cf. *Lenzi and Kowarsch 2021*). A second observation worth noting is that transparency efforts and value plurality sometimes conflict, for instance, when small modeling teams cannot provide the extensive documentation and data often demanded in transparency efforts. This results in the alternative value perspective appearing less credible and unable to fulfill transparency demands. This suggests there might be trade-offs between transparency and plurality or between different kinds of transparency.

This concludes this critical investigation into value aspects in integrated modeling. The focus on value judgments in mitigation pathways provided by IAMs and the shortcomings of current integrated modeling should, though, not conceal that the most relevant task is practical. IAM pathways consistently indicate steep emission reductions necessary to stay within the climate targets. For years, real-world emissions have risen or, at best, moved sideways. If we consider justice consideration, the curves shown to be necessary by the models would arguably be even steeper for the current generation and the Global North. Wherever they are precisely, though, currently the world is on the dangerous trajectory of largely ignoring these facts. We need a way to stay within the guardrails provided by the climate targets, crossing which' would throw humanity into a seriously more dangerous planetary future. Getting clearer on what the models tell us is necessary, and I hope this book provides some material for insight in this regard. However, humanity must turn this around quickly to avoid a gravely more dangerous world. If we fail to change, the modeled pathways from IAMs will be a testament to our collective moral failure. We will not be able to say we did not know how a world more just was possible.

References

Abend, Gabriel. 2019. "Thick Concepts and Sociological Research." *Sociological Theory* 37 (3): 209–33. https://doi.org/10.1177/0735275119869979.

Adler, Matthew D. 2011. *Well-Being and Fair Distribution Beyond Cost-Benefit Analysis.* Oxford University Press. https://doi.org/10.1093/acprof:oso/9780195384994.001.0001.

Adler, Matthew D. 2016. "Benefit–Cost Analysis and Distributional Weights: An Overview." *Review of Environmental Economics and Policy* 10 (2): 264–85. https://doi.org/10.1093/reep/rew005.

Adler, Matthew D. 2019. *Measuring Social Welfare: An Introduction.* New York: Oxford University Press.

Adler, Matthew D., and Nils Holtug. 2019. "Prioritarianism: A Response to Critics." *Politics, Philosophy & Economics* 18 (2): 101–44. https://doi.org/10.1177/1470594X19828022.

Adler, Matthew D., and Nicolas Treich. 2015. "Prioritarianism and Climate Change." *Environmental and Resource Economics* 62 (2): 279–308. https://doi.org/10.1007/s10640-015-9960-7.

Alexandrova, Anna. 2017. *A Philosophy for the Science of Well-Being.*

Allen, M. R., H. de Coninck, O. P. Dube, O. Hoegh-Guldberg, D. Jacob, K. Jiang, A. Revi, et al. 2018. "2018: Technical Summary." In *Global Warming of 1.5°C*, edited by Masson-Delmotte, V., P. Zhai, H.-O. Pörtner, D. Roberts, J. Skea, P.R. Shukla, et al. Cambridge, UK; New York, NY, USA: Cambridge University Press.

Anderson, Elizabeth. 2002. "Situated Knowledge and the Interplay of Value Judgments and Evidence in Scientific Inquiry." In *In the Scope of Logic, Methodology and Philosophy of Science*, edited by Peter Gärdenfors, Jan Woleński, and Katarzyna Kijania-Placek, 497–517. Dordrecht: Springer Netherlands. https://doi.org/10.1007/978-94-017-0475-5_8.

Anderson, Elizabeth. 2004. "Uses of Value Judgments in Science: A General Argument, with Lessons from a Case Study of Feminist Research on Divorce." *Hypatia* 19 (1): 1–24. https://doi.org/10.2979/hyp.2004.19.1.1.

Anderson, Elizabeth. 2009. "Democracy: Instrumental Vs. Non-Instrumental Value." In *Contemporary Debates in Political Philosophy*, edited by Thomas Christiano and John Philip Christman, 213–27. Contemporary Debates in Philosophy. Chichester, U.K.; Malden, MA: Wiley-Blackwell. https://doi.org/10.1002/9781444310399.ch12.

Anderson, Kevin. 2019. "Wrong Tool for the Job." *Nature* 573 (7774): 348. https://doi.org/10.1038/d41586-019-02744-9.

© The Editor(s) (if applicable) and The Author(s) 2025

S. Hollnaicher, *Assessing Feasibility with Value-laden Models*,

https://doi.org/10.1007/978-3-662-70714-2

Anderson, Kevin, and Glen Peters. 2016. "The Trouble with Negative Emissions." *Science (New York, N.Y.)* 354 (6309): 182–83. https://doi.org/10.1126/science.aah4567.

Anthoff, David, and Richard S. J. Tol. 2010. "On International Equity Weights and National Decision Making on Climate Change." *Journal of Environmental Economics and Management* 60 (1): 14–20. https://doi.org/10.1016/j.jeem.2010.04.002.

Aristoteles. 1985. *Nikomachische Ethik: 5.* Vol. 5. Philosophische Bibliothek. Hamburg: Meiner.

Atkinson, Giles, Ian J. Bateman, and Susana Mourato. 2014. "Valuing Ecosystem Services and Biodiversity." In *Nature in the Balance*, edited by Dieter Helm and Cameron Hepburn, 100–134. Oxford University Press. https://doi.org/10.1093/acprof:oso/9780199676880.003.0006.

Backman, Isabella, Marshall Burke, and Lawrence Goulder. 7.6.2021. "Stanford Explainer: Social Cost of Carbon," https://news.stanford.edu/2021/06/07/professors-explain-social-cost-carbon/.

Barker, Terry, and Douglas Crawford-Brown. 2013. "Are Estimated Costs of Stringent Mitigation Biased?" *Climatic Change* 121 (2): 129–38. https://doi.org/10.1007/s10584-013-0855-8.

Barker, Terry, and Katie Jenkins. 2007. "The Costs of Avoiding Dangerous Climate Change: Estimates Derived from a Meta-Analysis of the Literature." A briefing paper for the Human Development Report.

Barlas, Yaman, and Stanley Carpenter. 1990. "Philosophical Roots of Model Validation: Two Paradigms." *System Dynamics Review* 6 (2): 148–66. https://doi.org/10.1002/sdr.4260060203.

Barrotta, Pierluigi. 2018. "Values and Inductive Risk." In *Scientists, Democracy and Society*, edited by Pierluigi Barrotta, 16:49–82. Logic, Argumentation & Reasoning. Cham: Springer International Publishing. https://doi.org/10.1007/978-3-319-74938-9_3.

Beck, Marisa, and Tobias Krueger. 2016. "The Epistemic, Ethical, and Political Dimensions of Uncertainty in Integrated Assessment Modeling." *Wiley Interdisciplinary Reviews: Climate Change* 7 (5): 627–45. https://doi.org/10.1002/wcc.415.

Beck, Silke, and Martin Mahony. 2017. "The IPCC and the Politics of Anticipation." *Nature Climate Change* 7 (5): 311–13. https://doi.org/10.1038/nclimate3264.

Beck, Silke, and Martin Mahony. 2018a. "The IPCC and the New Map of Science and Politics." *Wiley Interdisciplinary Reviews: Climate Change* 9 (6): e547. https://doi.org/10.1002/wcc.547.

Beck, Silke, and Martin Mahony. 2018b. "The Politics of Anticipation: The IPCC and the Negative Emissions Technologies Experience." *Global Sustainability* 1. https://doi.org/10.1017/sus.2018.7.

Beck, Silke, and Jeroen Oomen. 2021. "Imagining the Corridor of Climate Mitigation – What Is at Stake in IPCC's Politics of Anticipation?" *Environmental Science & Policy* 123: 169–78. https://doi.org/10.1016/j.envsci.2021.05.011.

Beckerman, Wilfred. 2017. *Economics as Applied Ethics: Fact and Value in Economic Policy.* Second edition. Cham, Switzerland: Palgrave Macmillan. https://doi.org/10.1007/978-3-319-50319-6.

Beitz, Charles R. 2001. "Does Global Inequality Matter?" *Metaphilosophy* 32 (1/2): 95–112.

Bentham, Jeremy. 1907 [1789]. *An Introduction to the Principles of Morals and Legislation.* Oxford: Clarendon Press.

Berkey, Brian. 2021. "Climate Justice, Feasibility Constraints, and the Role of Political Philosophy." In *Climate Justice and Feasibility: Moral and Practical Concerns in a Warming World*, edited by Sarah Kenehan and Corey Katz. Rowman & Littlefield Publisher.

Betz, Gregor. 2013. "In Defence of the Value Free Ideal." *European Journal for Philosophy of Science* 3 (2): 207–20. https://doi.org/10.1007/s13194-012-0062-x.

Biddle, Justin B. 2013. "State of the Field: Transient Underdetermination and Values in Science." *Studies in History and Philosophy of Science Part A* 44 (1): 124–33. https://doi.org/10.1016/j.shpsa.2012.09.003.

Biddle, Justin B., and Anna Leuschner. 2015. "Climate Skepticism and the Manufacture of Doubt: Can Dissent in Science Be Epistemically Detrimental?" *European Journal for Philosophy of Science* 5 (3): 261–78. https://doi.org/10.1007/s13194-014-0101-x.

Bistline, John, Mark Budolfson, and Blake Francis. 2021. "Deepening Transparency about Value-Laden Assumptions in Energy and Environmental Modelling: Improving Best Practices for Both Modellers and Non-Modellers." *Climate Policy* 21 (1): 1–15. https://doi.org/10.1080/14693062.2020.1781048.

Blanchard, Elodie Vieille. 2010. "Modelling the Future: An Overview of the 'Limits to Growth' Debate." *Centaurus* 52 (2): 91–116. https://doi.org/10.1111/j.1600-0498.2010.00173.x.

Blesh, Jennifer, Lesli Hoey, Andrew D. Jones, Harriet Friedmann, and Ivette Perfecto. 2019. "Development Pathways Toward 'Zero Hunger'." *World Development* 118: 1–14. https://doi.org/10.1016/j.worlddev.2019.02.004.

Blum, Mareike. 2022. "Co-Producing Sustainability Research with Citizens: Empirical Insights from Co-Produced Problem Frames with Randomly Selected Citizens." *SSRN Electronic Journal*. https://doi.org/10.2139/ssrn.4220642.

Boran, Idil, and Kenneth Shockley. 2021. "Governance Towards Goals: Synergies, Equity, Feasibility." In *Climate Justice and Feasibility: Moral and Practical Concerns in a Warming World*, edited by Sarah Kenehan and Corey Katz, 35–58. Rowman & Littlefield Publisher.

Bosetti, Valentina, Emanuele Massetti, and Massimo Tavoni. 2007. "The WITCH Model: Structure, Baseline, Solutions." *SSRN Electronic Journal*. https://doi.org/10.2139/ssrn.960746.

Botzen, W. J. Wouter, and Jeroen C. J. M. van den Bergh. 2014. "Specifications of Social Welfare in Economic Studies of Climate Policy: Overview of Criteria and Related Policy Insights." *Environmental and Resource Economics* 58 (1): 1–33. https://doi.org/10.1007/s10640-013-9738-8.

Box, George E. P., and Norman Richard Draper. 1987. *Empirical Model-Building and Response Surfaces*. Wiley Series in Probability and Mathematical Statistics. New York: Wiley.

Brack, Duncan, and Richard King. 2020. *Net Zero and Beyond: What Role for Bioenergy with Carbon Capture and Storage?* London: Chatham House.

Brennan, Geoffrey. 2013. "Feasibility in Optimizing Ethics." *Social Philosophy and Policy* 30 (1–2): 314–29. https://doi.org/10.1017/S0265052513000150.

Brennan, Geoffrey, and Geoffrey Sayre-McCord. 2016. "Do Normative Facts Matter ... To What Is Feasible?" *Social Philosophy and Policy* 33 (1–2): 434–56. https://doi.org/10.1017/S0265052516000194.

Brennan, Geoffrey, and Nicholas Southwood. 2007. "Feasibility in Action and Attitude." In *Hommage à Wlodek*, edited by T. Rønnow-Rasmussen, B. Petersson, J. Josefsson, and D. Egonssson.

Broome, John. 1994. "Discounting the Future." *Philosophy and Public Affairs* 23 (2): 128–56.

Broome, John. 2012. *Climate Matters: Ethics in a Warming World*. Norton Global Ethics Series. WW Norton & Company.

Brown, Mark B. 2006. "Survey Article: Citizen Panels and the Concept of Representation*." *Journal of Political Philosophy* 14 (2): 203–25. https://doi.org/10.1111/j.1467-9760.2006.00245.x.

Brown, Mark B. 2018. "Deliberation and Representation." In *The Oxford Handbook of Deliberative Democracy*, edited by André Bächtiger, 170–86. Oxford Handbooks. Oxford: Oxford University Press. https://doi.org/10.1093/oxfordhb/9780198747369.013.58.

Brown, Matthew J. 2013. "Values in Science Beyond Underdetermination and Inductive Risk." *Philosophy of Science* 80 (5): 829–39. https://doi.org/10.1086/673720.

Brutschin, Elina, Silvia Pianta, Massimo Tavoni, Keywan Riahi, Valentina Bosetti, Giacomo Marangoni, and Bas van Ruijven. 2021. "A Multidimensional Feasibility Evaluation of Low-Carbon Scenarios." *Environmental Research Letters* 16 (6): 064069. https://doi.org/10.1088/1748-9326/abf0ce.

Buchanan, Allen E. 1985. *Ethics, Efficiency, and the Market*. Oxford: Clarendon Press.

Buchanan, Allen E. 2002. "Political Legitimacy and Democracy." *Ethics* 112 (4): 689–719. https://doi.org/10.1086/340313.

Buchanan, Allen E. 2003. *Justice, Legitimacy, and Self-Determination*. Oxford University PressOxford. https://doi.org/10.1093/0198295359.001.0001.

Budolfson, Mark. 2021. "Political Realism, Feasibility Wedges, and Opportunities for Collective Action on Climate Change." In *Philosophy and Climate Change*, edited by Mark Budolfson, Tristram McPherson, and David Plunkett, 323–45. Oxford University Press. https://doi.org/10.1093/oso/9780198796282.003.0015.

Budolfson, Mark, David Anthoff, Francis Dennig, Frank Errickson, Kevin Kuruc, Dean Spears, and Navroz K. Dubash. 2021. "Utilitarian Benchmarks for Emissions and Pledges Promote Equity, Climate and Development." *Nature Climate Change* 11 (10): 827–33. https://doi.org/10.1038/s41558-021-01130-6.

Budolfson, Mark, Francis Dennig, Marc Fleurbaey, Asher Siebert, and Robert H. Socolow. 2017. "The Comparative Importance for Optimal Climate Policy of Discounting, Inequalities and Catastrophes." *Climatic Change* 145 (3): 481–94. https://doi.org/10.1007/s10584-017-2094-x.

Butnar, Isabela, Oliver Broad, Baltazar Solano Rodriguez, and Paul E. Dodds. 2020. "The Role of Bioenergy for Global Deep Decarbonization: CO 2 Removal or Low–Carbon Energy?" *GCB Bioenergy* 12 (3): 198–212. https://doi.org/10.1111/gcbb.12666.

Byers, Edward, Volker Krey, Elmar Kriegler, Keywan Riahi, Roberto Schaeffer, Jarmo Kikstra, Robin Lamboll, et al. 2022. "AR6 Scenario Explorer and Database Hosted by IIASA: Database." https://doi.org/10.5281/zenodo.5886911.

Calvin, Katherine, Pralit Patel, Leon Clarke, Ghassem Asrar, Ben Bond-Lamberty, Ryna Yiyun Cui, Alan Di Vittorio, et al. 2019. "GCAM V5.1: Representing the Linkages Between Energy, Water, Land, Climate, and Economic Systems." *Geoscientific Model Development* 12 (2): 677–98. https://doi.org/10.5194/gmd-12-677-2019.

Caney, Simon. 2005. "Cosmopolitan Justice, Responsibility, and Global Climate Change." *Leiden Journal of International Law* 18 (4): 747–75. https://doi.org/10.1017/S09221565 05002992.

Caney, Simon. 2009. "Climate Change and the Future: Discounting for Time, Wealth, and Risk." *Journal of Social Philosophy* 40 (2): 163–86. https://doi.org/10.1111/j.1467-9833. 2009.01445.x.

Caney, Simon. 2010. "Climate Change and the Duties of the Advantaged." *Critical Review of International Social and Political Philosophy* 13 (1): 203–28. https://doi.org/10.1080/ 13698230903326331.

Caney, Simon. 2014a. "Climate Change, Intergenerational Equity and the Social Discount Rate." *Politics, Philosophy & Economics* 13 (4): 320–42. https://doi.org/10.1177/147059 4X14542566.

Caney, Simon. 2014b. "Two Kinds of Climate Justice: Avoiding Harm and Sharing Burdens." *Journal of Political Philosophy* 22 (2): 125–49. https://doi.org/10.1111/jopp.12030.

Caney, Simon. 2016. "Climate Change and Non-Ideal Theory." In *Climate Justice in a Non-Ideal World*, edited by Clare Heyward and Dominic Roser, 21–42. Oxford University Press. https://doi.org/10.1093/acprof:oso/9780198744047.003.0002.

Caney, Simon. 2018. "Distributive Justice and Climate Change." In *The Oxford Handbook of Distributive Justice*, edited by Serena Olsaretti, 664–88. Oxford, UK; New York, NY: Oxford University Press.

Carbon Brief. 2023. "Carbon Brief: Clear on Climate." https://www.carbonbrief.org/.

Carnap, Rudolf. 1945. "The Two Concepts of Probability: The Problem of Probability." *Philosophy and Phenomenological Research* 5 (4): 513. https://doi.org/10.2307/2102817.

Carnap, Rudolf. 1947. *Meaning and Necessity: A Study in Semantics and Modal Logic.* 2. ed., 1956. Chicago: University of Chicago Press.

Carnap, Rudolf. 1950. *Logical Foundations of Probability.* First edition. Chicago: University of Chicago Press.

Carrier, Martin. 2011. "Underdetermination as an Epistemological Test Tube: Expounding Hidden Values of the Scientific Community." *Synthese* 180 (2): 189–204. https://doi.org/ 10.1007/s11229-009-9597-6.

Carrier, Martin. 2013. "Values and Objectivity in Science: Value-Ladenness, Pluralism and the Epistemic Attitude." *Science & Education* 22 (10): 2547–68. https://doi.org/10.1007/ s11191-012-9481-5.

Carrier, Martin. 2019. "How to Conceive of Science for the Benefit of Society: Prospects of Responsible Research and Innovation." *Synthese.* https://doi.org/10.1007/s11229-019-02254-1.

Carrier, Martin. 2021. "What Does Good Science-Based Advice to Politics Look Like?" *Journal for General Philosophy of Science*, no. 53: 5–21. https://doi.org/10.1007/s10838-021-09574-2.

Carrol, Lewis. 1894 [2015]. *Sylvie and Bruno Concluded.* Project Gutenberg EBook. https://www.gutenberg.org/files/48795/48795-h/48795-h.htm.

Cartwright, Nancy. 2006. "Well-Ordered Science: Evidence for Use." *Philosophy of Science* 73 (5): 981–90. https://doi.org/10.1086/518803.

Chang, Hasok. 2012. "Pluralism in Science: A Call to Action." In *Is Water H2O?*, edited by Hasok Chang, 293:253–301. Boston Studies in the Philosophy of Science. Dordrecht: Springer Netherlands. https://doi.org/10.1007/978-94-007-3932-1_5.

Charney, Jule A., and et al. 1979. *Carbon Dioxide and Climate: A Scientific Assessment.* Washington, District of Columbia: National Academy of Sciences. https://doi.org/10. 17226/12181.

Chen, Xiaotong, Fang Yang, Shining Zhang, Behnam Zakeri, Xing Chen, Changyi Liu, and Fangxin Hou. 2021. "Regional Emission Pathways, Energy Transition Paths and Cost Analysis Under Various Effort-Sharing Approaches for Meeting Paris Agreement Goals." *Energy* 232: 121024. https://doi.org/10.1016/j.energy.2021.121024.

Christiano, Thomas. 2004. "The Authority of Democracy." *Journal of Political Philosophy* 12 (3): 266–90. https://doi.org/10.1111/j.1467-9760.2004.00200.x.

Christiano, Thomas. 2008. *The Constitution of Equality: Democratic Authority and Its Limits.* Oxford; New York: Oxford Univ. Press.

Churchman, C. West. 1948. "Statistics, Pragmatics, Induction." *Philosophy of Science* 15 (3): 249–68.

Clarke, L., K. Jiang, K. Akimoto, M. Babiker, G. Blanford, K. Fisher-Vanden, J.-C. Hourcade, et al. 2014. "2014: Assessing Transformation Pathways." In *Climate Change 2014: Mitigation of Climate Change*, edited by Edenhofer, O., R. Pichs-Madruga, Y. Sokona, E. Farahani, S. Kadner, K. Seyboth, A. Adler, I. Baum, S. Brunner, P. Eickemeier, B. Kriemann, J. Savolainen, S. Schlömer, C. von Stechow, T. Zwickel and J.C. Minx, 413–510. Cambridge, UK & New York, NY, USA: Cambridge University Press.

Climate Action Tracker. 2019a. "December 2019 Global Update." Berlin.

Climate Action Tracker. 2019b. "The CAT Thermometer." https://climateactiontracker.org/global/cat-thermometer/.

Cohen, Gerald A. 2009. *Why Not Socialism?* Princeton, NJ: Princeton University Press.

Collins English Dictionary. 2023. "Definition of 'Feasible'." https://www.collinsdictionary.com/dictionary/english/feasible.

Creutzig, Felix. 2016. "Economic and Ecological Views on Climate Change Mitigation with Bioenergy and Negative Emissions." *GCB Bioenergy* 8 (1): 4–10. https://doi.org/10.1111/gcbb.12235.

Creutzig, Felix, Peter Agoston, Jan Christoph Goldschmidt, Gunnar Luderer, Gregory Nemet, and Robert C. Pietzcker. 2017. "The Underestimated Potential of Solar Energy to Mitigate Climate Change." *Nature Energy* 2 (9). https://doi.org/10.1038/nenergy.2017.140.

Creutzig, Felix, N. H. Ravindranath, Göran Berndes, Simon Bolwig, Ryan Bright, Francesco Cherubini, Helena Chum, et al. 2015. "Bioenergy and Climate Change Mitigation: An Assessment." *GCB Bioenergy* 7 (5): 916–44. https://doi.org/10.1111/gcbb.12205.

Crisp, Roger. 2003. "Equality, Priority, and Compassion." *Ethics* 113 (4): 745–63. https://doi.org/10.1086/373954.

Crisp, Roger. 2023. "Well-Being." In *The Stanford Encyclopedia of Philosophy*, edited by Edward N. Zalta. Metaphysics Research Lab, Stanford University. https://plato.stanford.edu/archives/win2021/entries/well-being/.

Daioglou, Vassilis, Steven K. Rose, Nico Bauer, Alban Kitous, Matteo Muratori, Fuminori Sano, Shinichiro Fujimori, et al. 2020. "Bioenergy Technologies in Long-Run Climate Change Mitigation: Results from the EMF-33 Study." *Climatic Change* 163 (3): 1603–20. https://doi.org/10.1007/s10584-020-02799-y.

Daioglou, Vassilis, Bas J. van Ruijven, and Detlef P. van Vuuren. 2012. "Model Projections for Household Energy Use in Developing Countries." *Energy* 37 (1): 601–15. https://doi. org/10.1016/j.energy.2011.10.044.

Dasgupta, Partha. 2008. "Discounting Climate Change." *Journal of Risk and Uncertainty* 37 (2–3): 141–69. https://doi.org/10.1007/s11166-008-9049-6.

Deng, Yvonne Y., Martin Haigh, Willemijn Pouwels, Lou Ramaekers, Ruut Brandsma, Sven Schimschar, Jan Grözinger, and David de Jager. 2015. "Quantifying a Realistic, Worldwide Wind and Solar Electricity Supply." *Global Environmental Change* 31: 239–52. https://doi.org/10.1016/j.gloenvcha.2015.01.005.

Dennig, Francis. 2018. "Climate Change and the Re-Evaluation of Cost-Benefit Analysis." *Climatic Change* 151 (1): 43–54. https://doi.org/10.1007/s10584-017-2047-4.

De-Shalit, Avner. 1995. *Why Posterity Matters: Environmental Policies and Future Generations.* 1. publ. Environmental Philosophies Series. London: Routledge.

Després, J., K. Keramidas, A. Kitous, and A. Schmitz. 2017. *POLES-JRC Model Documentation.* Vol. 28728. EUR, Scientific and Technical Research Series. Luxembourg: Publications Office of the European Union.

Dewey, John. 2008. "Logic: The Theory of Inquiry." In *The Collected Works of John Dewey*, edited by John Dewey. Vol. 12. Carbondale: Southern Illinois Univ. Press.

Dietz, Simon, C. Hepburn, and N. Stern. 2009. "Economics, Ethics and Climate Change." *IOP Conference Series: Earth and Environmental Science* 6 (12): 122003. https://doi.org/10.1088/1755-1307/6/2/122003.

Djordjevic, Charles, and Catherine Herfeld. 2021. "Thick Concepts in Economics: The Case of Becker and Murphy's Theory of Rational Addiction." *Philosophy of the Social Sciences* 51 (4): 371–99. https://doi.org/10.1177/00483931211008541.

Dooley, Kate, Christian Holz, Sivan Kartha, Sonja Klinsky, J. Timmons Roberts, Henry Shue, Harald Winkler, et al. 2021. "Ethical Choices Behind Quantifications of Fair Contributions Under the Paris Agreement." *Nature Climate Change* 11 (4): 300–305. https://doi.org/10.1038/s41558-021-01015-8.

Douglas, Heather. 2000. "Inductive Risk and Values in Science." *Philosophy of Science* 67 (4): 559–79. https://doi.org/10.1086/392855.

Douglas, Heather. 2004. "The Irreducible Complexity of Objectivity." *Synthese* 138 (3): 453–73. https://doi.org/10.1023/B:SYNT.0000016451.18182.91.

Douglas, Heather. 2009. *Science, Policy, and the Value-Free Ideal.* University of Pittsburgh Press. https://doi.org/10.2307/j.ctt6wrc78.

Douglas, Heather. 2015. "Values in Science." In *The Oxford Handbook of Philosophy of Science*, edited by Paul Humphreys. Oxford University Press.

Doyal, Len, and Ian Gough. 1991. *A Theory of Human Need.* London: Macmillan.

Drupp, Moritz A., Mark C. Freeman, Ben Groom, and Frikk Nesje. 2018. "Discounting Disentangled." *American Economic Journal: Economic Policy* 10 (4): 109–34. https://doi.org/10.1257/pol.20160240.

Dryzek, John S. 2010. *Foundations and Frontiers of Deliberative Governance.* Oxford: Oxford University Press.

Dryzek, John S., and Jonathan Pickering. 2019. *The Politics of the Anthropocene.* First edition. Oxford: Oxford University Press.

Du Robiou Pont, Yann, M. Louise Jeffery, Johannes Gütschow, Joeri Rogelj, Peter Christoff, and Malte Meinshausen. 2017. "Equitable Mitigation to Achieve the Paris Agreement Goals." *Nature Climate Change* 7 (1): 38–43. https://doi.org/10.1038/nclimate3186.

Dupré, John. 2007. "Fact and Value." In *Value-Free Science?*, edited by Harold Kincaid, John Dupré, and Alison Wylie, 27–41. Oxford: Oxford University Press. https://doi.org/10.1093/acprof:oso/9780195308969.003.0003.

Dworkin, Ronald. 1981. "What Is Equality? Part 2: Equality of Resources." *Philosophy & Public Affairs* 10 (4): 283–345.

Dyke, James, Robert Watson, and Wolfgang Knorr. 2021. "Climate Scientists: Concept of Net Zero Is a Dangerous Trap." https://theconversation.com/climate-scientists-concept-of-net-zero-is-a-dangerous-trap-157368.

Edenhofer, Ottmar, and Martin Kowarsch. 2015. "Cartography of Pathways: A New Model for Environmental Policy Assessments." *Environmental Science & Policy* 51: 56–64. https://doi.org/10.1016/j.envsci.2015.03.017.

Edenhofer, Ottmar, and Jan Minx. 2014. "Climate Policy. Mapmakers and Navigators, Facts and Values." *Science (New York, N.Y.)* 345 (6192): 37–38. https://doi.org/10.1126/science.1255998.

Eigi, Jaana. 2019. "How to Think about Shared Norms and Pluralism Without Circularity: A Reply to Anna Leuschner." *Studies in History and Philosophy of Science Part A* 75: 51–56. https://doi.org/10.1016/j.shpsa.2019.01.007.

Elliott, Kevin C. 2011. *Is a Little Pollution Good for You? Incorporating Societal Values in Environmental Research*. Environmental Ethics and Science Policy Series. Oxford: Oxford University Press.

Elliott, Kevin C. 2017. *A Tapestry of Values: An Introduction to Values in Science*. New York, NY: Oxford University Press.

Elliott, Kevin C. 2020. "A Taxonomy of Transparency in Science." *Canadian Journal of Philosophy*, 1–14. https://doi.org/10.1017/can.2020.21.

Elliott, Kevin C. 2021. "The Value-Ladenness of Transparency in Science: Lessons from Lyme Disease." *Studies in History and Philosophy of Science Part A* 88: 1–9. https://doi.org/10.1016/j.shpsa.2021.03.008.

Elliott, Kevin C., and David B. Resnik. 2014. "Science, Policy, and the Transparency of Values." *Environmental Health Perspectives* 122 (7): 647–50. https://doi.org/10.1289/ehp.1408107.

Emmerling, Johannes, Laurent Drouet, Kaj-Ivar van der Wijst, Detlef van Vuuren, Valentina Bosetti, and Massimo Tavoni. 2019. "The Role of the Discount Rate for Emission Pathways and Negative Emissions." *Environmental Research Letters* 14 (10): 104008. https://doi.org/10.1088/1748-9326/ab3cc9.

Erman, Eva, and Niklas Möller. 2020. "A World of Possibilities: The Place of Feasibility in Political Theory." *Res Publica* 26 (1): 1–23. https://doi.org/10.1007/s11158-018-09415-y.

Ernst, Gerhard. 2008. *Die Objektivität Der Moral: Teilw. Zugl.: München, Univ., Habil.-Schr., 2004 u.d.t.: Ernst, Gerhard: Die Natur Der Moral*. Paderborn: Mentis.

Estlund, David. 2020. *Utopophobia: On the Limits (If Any) of Political Philosophy*. Princeton: Princeton University Press. https://doi.org/10.1515/9780691197500.

Estlund, David M. 2008. *Democratic Authority: A Philosophical Framework*. Princeton, NJ: Princeton University Press.

Evans, Simon, and Zeke Hausfather. 2018. "Q&a: How 'Integrated Assessment Models' Are Used to Study Climate Change." https://www.carbonbrief.org/qa-how-integrated-assessment-models-are-used-to-study-climate-change.

Fajardy, Mathilde, Alexandre Köberle, Niall Mac Dowell, and Andrea Fantuzzi. 2019. "BECCS Deployment: A Reality Check." *Grantham Institute Briefing Paper*, no. 28: 1–14.

Fajardy, Mathilde, and Niall Mac Dowell. 2017. "Can BECCS Deliver Sustainable and Resource Efficient Negative Emissions?" *Energy & Environmental Science* 10 (6): 1389–1426. https://doi.org/10.1039/C7EE00465F.

Fearon, James D. 1999. "Electoral Accountability and the Control of Politicians: Selecting Good Types Versus Sanctioning Poor Performance." In *Democracy, Accountability, and Representation*, edited by Adam Przeworski, Susan Carol Stokes, and Bernard Manin, 55–97. Cambridge Studies in the Theory of Democracy. Cambridge, U.K.; New York: Cambridge University Press. https://doi.org/10.1017/CBO9781139175104.003.

Feest, Uljana. 2005. "Operationism in Psychology: What the Debate Is about, What the Debate Should Be About." *Journal of the History of the Behavioral Sciences* 41 (2): 131–49. https://doi.org/10.1002/jhbs.20079.

Fisher, B. S., N. Nakicenovic, K. Alfsen, J. Corfee Morlot, F. de La Chesnaye, J.-Ch Hourcade, K. Jiang, et al. 2007. "2007: Issues Related to Mitigation in the Long Term Context." In *Climate Change 2007*. Cambridge, Mass.: Cambridge University Press.

Fishkin, James S. 1991. *Democracy and Deliberation: New Directions for Democratic Reform.* New Haven: Yale Univ. Press.

Fishkin, James S. 1997. *The Voice of the People: Public Opinion and Democracy.* [New ed.]. New Haven: Yale University Press.

Fleurbaey, Marc, and Stéphane Zuber. 2012. "Climate Policies Deserve a Negative Discount Rate." *Chicago Journal of International Law* 13 (2): 565–95.

Fourie, Carina. 2016. "The Sufficiency View." In *What Is Enough?*, edited by Carina Fourie and Annette Rid, 11–29. Oxford University Press. https://doi.org/10.1093/acprof:oso/9780199385263.003.0002.

Frank, David M. 2019. "Ethics of the Scientist Qua Policy Advisor: Inductive Risk, Uncertainty, and Catastrophe in Climate Economics." *Synthese* 196 (8): 3123–38. https://doi.org/10.1007/s11229-017-1617-3.

Fridahl, Mathias, and Mariliis Lehtveer. 2018. "Bioenergy with Carbon Capture and Storage (BECCS): Global Potential, Investment Preferences, and Deployment Barriers." *Energy Research & Social Science* 42: 155–65. https://doi.org/10.1016/j.erss.2018.03.019.

Frisch, Mathias. 2013. "Modeling Climate Policies: A Critical Look at Integrated Assessment Models." *Philosophy & Technology* 26 (2): 117–37. https://doi.org/10.1007/s13347-013-0099-6.

Frisch, Mathias. 2017. "Climate Policy in the Age of Trump." *Kennedy Institute of Ethics Journal* 27 (2S): E-87-E-106. https://doi.org/10.1353/ken.2017.0027.

Frisch, Mathias. 2018. "Modeling Climate Policies: The Social Cost of Carbon and Uncertainties in Climate Predictions." In *Climate Modelling*, edited by Elisabeth Anne Lloyd and Eric B. Winsberg, 413–49. Basingstoke, Hampshire: Palgrave Macmillan.

Fritsch, Matthias. 2019. "All-Affected Principle." In *The Cambridge Habermas Lexicon*, edited by Amy Allen, 27:7–8. Cambridge; New York: Cambridge University Press. https://doi.org/10.1017/9781316771303.004.

Fuhrman, Jay, Andres Clarens, Katherine Calvin, Scott C. Doney, James A. Edmonds, Patrick O'Rourke, Pralit Patel, Shreekar Pradhan, William Shobe, and Haewon McJeon. 2021. "The Role of Direct Air Capture and Negative Emissions Technologies in the Shared Socioeconomic Pathways Towards +1.5 °c and +2 °c Futures." *Environmental Research Letters* 16 (11): 114012. https://doi.org/10.1088/1748-9326/ac2db0.

Fujimori, Shinichiro, Hancheng Dai, Toshihiko Masui, and Yuzuru Matsuoka. 2016. "Global Energy Model Hindcasting." *Energy* 114: 293–301. https://doi.org/10.1016/j.energy.2016.08.008.

Fuss, Sabine, Josep G. Canadell, Glen P. Peters, Massimo Tavoni, Robbie M. Andrew, Philippe Ciais, Robert B. Jackson, et al. 2014. "Betting on Negative Emissions." *Nature Climate Change* 4 (10): 850–53. https://doi.org/10.1038/nclimate2392.

Fuss, Sabine, William F. Lamb, Max W. Callaghan, Jérôme Hilaire, Felix Creutzig, Thorben Amann, Tim Beringer, et al. 2018. "Negative Emissions – Part 2: Costs, Potentials and Side Effects." *Environmental Research Letters* 13 (6): 063002. https://doi.org/10.1088/1748-9326/aabf9f.

Fyson, Claire L., Susanne Baur, Matthew Gidden, and Carl-Friedrich Schleussner. 2020. "Fair-Share Carbon Dioxide Removal Increases Major Emitter Responsibility." *Nature Climate Change* 10 (9): 836–41. https://doi.org/10.1038/s41558-020-0857-2.

Gambhir, Ajay, Laurent Drouet, David McCollum, Tamaryn Napp, Dan Bernie, Adam Hawkes, Oliver Fricko, et al. 2017. "Assessing the Feasibility of Global Long-Term Mitigation Scenarios." *Energies* 10 (1): 89. https://doi.org/10.3390/en10010089.

Gambhir, Ajay, Gaurav Ganguly, and Shivika Mittal. 2022. "Climate Change Mitigation Scenario Databases Should Incorporate More Non-IAM Pathways." *Joule* 6 (12): 2663–67. https://doi.org/10.1016/j.joule.2022.11.007.

Gambhir, Ajay, and Massimo Tavoni. 2019. "Direct Air Carbon Capture and Sequestration: How It Works and How It Could Contribute to Climate-Change Mitigation." *One Earth* 1 (4): 405–9. https://doi.org/10.1016/j.oneear.2019.11.006.

Ganti, Gaurav, Matthew J. Gidden, Christopher J. Smith, Claire Fyson, Alexander Nauels, Keywan Riahi, and Carl-Friedrich Schleußner. 2023. "Uncompensated Claims to Fair Emission Space Risk Putting Paris Agreement Goals Out of Reach." *Environmental Research Letters* 18 (2): 024040. https://doi.org/10.1088/1748-9326/acb502.

Garard, Jennifer, Larissa Koch, and Martin Kowarsch. 2018. "Elements of Success in Multi-Stakeholder Deliberation Platforms." *Palgrave Communications* 4 (1). https://doi.org/10.1057/s41599-018-0183-8.

Gardiner, Stephen M. 2006. "A Perfect Moral Storm: Climate Change, Intergenerational Ethics and the Problem of Moral Corruption." *Environmental Values* 15 (3): 397–413.

Gardiner, Stephen M. 2011. *A Perfect Moral Storm: The Ethical Tragedy of Climate Change.* Environmental Ethics and Science Policy Series. Oxford: Oxford University Press.

Geden, Oliver. 2015. "Policy: Climate Advisers Must Maintain Integrity." *Nature* 521 (7550): 27–28. https://doi.org/10.1038/521027a.

Gieryn, Thomas F. 1983. "Boundary-Work and the Demarcation of Science from Non-Science: Strains and Interests in Professional Ideologies of Scientists." *American Sociological Review* 48 (6): 781. https://doi.org/10.2307/2095325.

Gilabert, Pablo. 2012. "Comparative Assessments of Justice, Political Feasibility, and Ideal Theory." *Ethical Theory and Moral Practice* 15 (1): 39–56. https://doi.org/10.1007/s10677-011-9279-6.

Gilabert, Pablo. 2017. "Justice and Feasibility." In *Political Utopias*, edited by Michael Weber and Kevin Vallier, 95–126. Oxford University Press. https://doi.org/10.1093/acp rof:oso/9780190280598.003.0006.

Gilabert, Pablo, and Holly Lawford-Smith. 2012. "Political Feasibility: A Conceptual Exploration." *Political Studies* 60 (4): 809–25. https://doi.org/10.1111/j.1467-9248.2011.009 36.x.

Google Books. 2023. "Ngram Viewer." https://books.google.com/ngrams.

Gosseries, Axel. 2016. "Sufficientarianism." In *Routledge Encyclopedia of Philosophy*. London: Routledge. https://doi.org/10.4324/9780415249126-S112-1.

Grant, Neil, Adam Hawkes, Shivika Mittal, and Ajay Gambhir. 2021. "Confronting Mitigation Deterrence in Low-Carbon Scenarios." *Environmental Research Letters* 16 (6): 064099. https://doi.org/10.1088/1748-9326/ac0749.

Grant, Neil, Adam Hawkes, Tamaryn Napp, and Ajay Gambhir. 2020. "The Appropriate Use of Reference Scenarios in Mitigation Analysis." *Nature Climate Change* 10 (7): 605–10. https://doi.org/10.1038/s41558-020-0826-9.

Grubler, Arnulf, Charlie Wilson, Nuno Bento, Benigna Boza-Kiss, Volker Krey, David L. McCollum, Narasimha D. Rao, et al. 2018. "A Low Energy Demand Scenario for Meeting the 1.5 °c Target and Sustainable Development Goals Without Negative Emission Technologies." *Nature Energy* 3 (6): 515–27. https://doi.org/10.1038/s41560-018-0172-6.

Guillemot, Hélène. 2017. "The Necessary and Inaccessible 1.5°c Objective: A Turning Point in the Relations Between Climate Science and Politics?" In *Globalising the Climate*, edited by Stefan Cihan Aykut, Jean Foyer, and Édouard Morena. Routledge Advances in Climate Change Research. London; New York: Routledge Taylor & Francis Group.

Guillery, Daniel. 2021. "The Concept of Feasibility: A Multivocal Account." *Res Publica* 27 (3): 491–507. https://doi.org/10.1007/s11158-020-09497-7.

Guivarch, C, E. Kriegler, J. Portugal-Pereira, V. Bosetti, J. Edmonds, M. Fischedick, et al. 2022. "Annex III: Scenarios and Modelling Methods." In *IPCC, 2022: Climate Change 2022: Mitigation of Climate Change. Contribution of Working Group III to the Sixth Assessment Report of the Intergovernmental Panel on Climate Change*. Cambridge, UK; New York, NY, USA: Cambridge University Press.

Gundersen, Torbjørn. 2020. "Value-Free yet Policy-Relevant? The Normative Views of Climate Scientists and Their Bearing on Philosophy." *Perspectives on Science* 28 (1): 89–118. https://doi.org/10.1162/posc_a_00334.

Habermas, Jürgen. 2019 [1992]. *Faktizität Und Geltung: Beiträge Zur Diskurstheorie Des Rechts Und Des Demokratischen Rechtsstaats*. 7. Auflage. Frankfurt am Main: Suhrkamp.

Haikola, Simon, Anders Hansson, and Mathias Fridahl. 2018. "Views of BECCS Among Modelers and Policymakers." In *Bioenergy with Carbon Capture and Storage from Global Potentials to Domestic Realities*, edited by Mathias Fridahl, 17–29. Brussels, Belgium: European Liberal Forum asbl.

Haikola, Simon, Anders Hansson, and Mathias Fridahl. 2019. "Map-Makers and Navigators of Politicised Terrain: Expert Understandings of Epistemological Uncertainty in Integrated Assessment Modelling of Bioenergy with Carbon Capture and Storage." *Futures* 114: 102472. https://doi.org/10.1016/j.futures.2019.102472.

Hamlin, Alan. 2017. "Feasibility Four Ways." *Social Philosophy and Policy* 34 (1): 209–31. https://doi.org/10.1017/S0265052517000103.

Hänsel, Martin C., Moritz A. Drupp, Daniel J. A. Johansson, Frikk Nesje, Christian Azar, Mark C. Freeman, Ben Groom, and Thomas Sterner. 2020. "Climate Economics Support for the UN Climate Targets." *Nature Climate Change* 10 (8): 781–89. https://doi.org/10. 1038/s41558-020-0833-x.

Hansson, Anders, Jonas Anshelm, Mathias Fridahl, and Simon Haikola. 2021. "Boundary Work and Interpretations in the IPCC Review Process of the Role of Bioenergy with Carbon Capture and Storage (BECCS) in Limiting Global Warming to 1.5°c." *Frontiers in Climate* 3. https://doi.org/10.3389/fclim.2021.643224.

Hare, Bill, Robert Brecha, and Michiel Schaeffer. 2018. "Integrated Assessment Models: What Are They and How Do They Arrive at Their Conclusions?" Edited by Climate Analytics. Policy Briefing. https://climateanalytics.org/media/climate_analytics_iams_b riefing_oct2018.pdf.

Harman, Elizabeth. 2015. "The Irrelevance of Moral Uncertainty." In *Oxford Studies in Metaethics, Volume 10*, edited by Russ Shafer-Landau, 53–79. Oxford University Press. https://doi.org/10.1093/acprof:oso/9780198738695.003.0003.

Harper, Anna B., Tom Powell, Peter M. Cox, Joanna House, Chris Huntingford, Timothy M. Lenton, Stephen Sitch, et al. 2018. "Land-Use Emissions Play a Critical Role in Land-Based Mitigation for Paris Climate Targets." *Nature Communications* 9 (1): 2938. https:// doi.org/10.1038/s41467-018-05340-z.

Hausfather, Zeke, and Glen P. Peters. 2020. "Emissions—the 'Business as Usual' Story Is Misleading." *Nature* 577 (7792): 618–20. https://doi.org/10.1038/d41586-020-00177-3.

Hausman, Daniel M., and Michael S. McPherson. 2006. *Economic Analysis, Moral Philosophy, and Public Policy*. Cambridge: Cambridge University Press. https://doi.org/10.1017/ CBO9780511754289.

Havstad, Joyce C., and Matthew J. Brown. 2017a. "Inductive Risk, Deferred Decisions, and Climate Science Advising." In *Exploring Inductive Risk*, edited by Kevin C. Elliott and Ted Richards, 101–26. Oxford: Oxford University Press. https://doi.org/10.1093/acprof: oso/9780190467715.003.0006.

Havstad, Joyce C., and Matthew J. Brown. 2017b. "Neutrality, Relevance, Prescription, and the IPCC." *Public Affairs Quarterly* 31 (4): 303–24.

Hawkins, Ed, and Rowan Sutton. 2009. "The Potential to Narrow Uncertainty in Regional Climate Predictions." *Bulletin of the American Meteorological Society* 90 (8): 1095–1108. https://doi.org/10.1175/2009BAMS2607.1.

Hellewell, Joel, Sam Abbott, Amy Gimma, Nikos I. Bosse, Christopher I. Jarvis, Timothy W. Russell, James D. Munday, et al. 2020. "Feasibility of Controlling COVID-19 Outbreaks by Isolation of Cases and Contacts." *The Lancet. Global Health* 8 (4): e488–96. https:// doi.org/10.1016/S2214-109X(20)30074-7.

Heyward, Clare. 2012. "Review of Climate Change Justice by Eric a. Posner, David Weisbach." *Carbon & Climate Law Review* 6 (1): 92–94.

Heyward, Clare, and Jörgen Ödalen. 2016. "A Free Movement Passport for the Territorially Dispossessed." In *Climate Justice in a Non-Ideal World*, edited by Clare Heyward and Dominic Roser, 208–26. Oxford University Press. https://doi.org/10.1093/acprof:oso/978 0198744047.003.0011.

Hickel, Jason, and Aljosa Slamersak. 2022. "Existing Climate Mitigation Scenarios Perpetuate Colonial Inequalities." *The Lancet. Planetary Health* 6 (7): e628–31. https://doi.org/ 10.1016/S2542-5196(22)00092-4.

Hilligardt, Hannah. 2022. "Looking Beyond Values: The Legitimacy of Social Perspectives, Opinions and Interests in Science." *European Journal for Philosophy of Science* 12 (4): 58. https://doi.org/10.1007/s13194-022-00490-w.

Hirschl, Bernd, Uwe Schwarz, Julika Weiß, Raoul Hirschberg, and Lukas Torliene. 2021. "Berlin Paris-Konform Machen: Eine Aktualisierung Der Machbarkeitsstudie Klimaneutrales Berlin 2050 Mit Blick Auf Die Anforderungen Aus Dem UN-Abkommen von Paris: Im Auftrag Des Landes Berlin, Vertreten Durch Die Senatsverwaltung für Umwelt, Verkehr Und Klimaschutz; Berlin." Berlin: Institut für ökologische Wirtschaftsforschung.

Hollnaicher, Simon. 2022. "On Economic Modeling of Carbon Dioxide Removal: Values, Bias, and Norms for Good Policy-Advising Modeling." *Global Sustainability*, 1–33. https://doi.org/10.1017/sus.2022.16.

Holman, Bennett, and Torsten Wilholt. 2022. "The New Demarcation Problem." *Studies in History and Philosophy of Science Part A* 91: 211–20. https://doi.org/10.1016/j.shpsa.2021.11.011.

Holz, Christian, Sivan Kartha, and Tom Athanasiou. 2018. "Fairly Sharing 1.5: National Fair Shares of a 1.5 °c-Compliant Global Mitigation Effort." *International Environmental Agreements: Politics, Law and Economics* 18 (1): 117–34. https://doi.org/10.1007/s10784-017-9371-z.

Höök, Mikael, Junchen Li, Kersti Johansson, and Simon Snowden. 2012. "Growth Rates of Global Energy Systems and Future Outlooks." *Natural Resources Research* 21 (1): 23–41. https://doi.org/10.1007/s11053-011-9162-0.

Houston, Jared. 2021. "The 'Pathway Problem,' Probabilistic Feasibility, and Non-Ideal Climate Justice." In *Climate Justice and Feasibility: Moral and Practical Concerns in a Warming World*, edited by Sarah Kenehan and Corey Katz. Rowman & Littlefield Publisher.

Hulme, Mike. 2016. "1.5 °c and Climate Research After the Paris Agreement." *Nature Climate Change* 6 (3): 222–24. https://doi.org/10.1038/nclimate2939.

Hume, David. 1739--40 [2001]. *A Treatise of Human Nature, (Eds.),*. Oxford: Oxford University Press.

IAMC. 2021. "IAMC Wiki: The Common Integrated Assessment Model (IAM) Documentation." https://www.iamcdocumentation.eu/index.php/IAMC_wiki.

IAMC. 2022. "Model Documentation—REMIND-MAgPIE." https://www.iamcdocumentation.eu/index.php/Model_Documentation_-_REMIND-MAgPIE.

Intemann, Kristen. 2015. "Distinguishing Between Legitimate and Illegitimate Values in Climate Modeling." *European Journal for Philosophy of Science* 5 (2): 217–32. https://doi.org/10.1007/s13194-014-0105-6.

IPCC. n. d. "What Is the IPCC?" https://www.ipcc.ch/languages-2/english/.

IPCC. 1990. "FAR Climate Change: Impacts Assessment of Climate Change." Edited by J. T. Houghton, G. J. Jenkins, and J. J. Ephraums. Cambridge: Cambridge University Press.

IPCC. 2014a. *Climate Change 2014: Impacts, Adaptation, and Vulnerability : Working Group II Contribution to the Fifth Assessment Report of the Intergovernmental Panel on Climate Change.* Cambridge: Cambridge University Press.

IPCC. 2014b. *Climate Change 2014: Mitigation of Climate Change: Contribution of Working Group III to the Fifth Assessment Report of the Intergovernmental Panel on Climate Change.* Cambridge, UK & New York, NY, USA: IPCC; Cambridge University Press.

IPCC. 2022. "IPCC, 2022: Climate Change 2022: Mitigation of Climate Change. Contribution of Working Group III to the Sixth Assessment Report of the Intergovernmental Panel on Climate Change." Cambridge, UK; New York, NY, USA: Cambridge University Press. https://doi.org/10.1017/9781009157926.

IPCC. 2023. "Synthesis Report of the IPCC Sixth Assessment Report: Summary for Policymakers." Edited by Hoesung et al. Lee.

Jabbour, Jason, and Christian Flachsland. 2017. "40 Years of Global Environmental Assessments: A Retrospective Analysis." *Environmental Science & Policy* 77: 193–202. https://doi.org/10.1016/j.envsci.2017.05.001.

Jasanoff, Sheila, ed. 2004. *States of Knowledge: The Co-Production of Science and Social Order*. International Library of Sociology. London; New York: Routledge.

Jeffrey, Richard C. 1956. "Valuation and Acceptance of Scientific Hypotheses." *Philosophy of Science* 23 (3): 237–46. https://doi.org/10.1086/287489.

Jewell, Jessica, and Aleh Cherp. 2020. "On the Political Feasibility of Climate Change Mitigation Pathways: Is It Too Late to Keep Warming Below 1.5°c?" *WIREs Climate Change* 11 (1): 19. https://doi.org/10.1002/wcc.621.

John, Stephen. 2015. "The Example of the IPCC Does Not Vindicate the Value Free Ideal: A Reply to Gregor Betz." *European Journal for Philosophy of Science* 5 (1): 1–13. https://doi.org/10.1007/s13194-014-0095-4.

John, Stephen. 2018. "Epistemic Trust and the Ethics of Science Communication: Against Transparency, Openness, Sincerity and Honesty." *Social Epistemology* 32 (2): 75–87. https://doi.org/10.1080/02691728.2017.1410864.

Kant, Immanuel. 1786 [2007]. *Grundlegung Zur Metaphysik Der Sitten: [GMS]*. Frankfurt am Main: Suhrkamp.

Kant, Immanuel. 1995. *Kritik Der Reinen Vernunft*. Suhrkamp-Taschenbuch Wissenschaft. Frankfurt am Main: Suhrkamp.

Kant, Immanuel, and Allen W. Wood. 1996. "On the Common Saying: That May Be Correct in Theory, but It Is of No Use in Practice (1793)." In *Immanuel Kant: Practical Philosophy*, edited by Mary J. Gregor, Immanuel Kant, and Allen W. Wood, 273–310. Cambridge: Cambridge University Press. https://doi.org/10.1017/CBO9780511813306.011.

Keen, Steve. 2021. "The Appallingly Bad Neoclassical Economics of Climate Change." *Globalizations* 18 (7): 1149–77. https://doi.org/10.1080/14747731.2020.1807856.

Keen, Steve, Timothy M. Lenton, Antoine Godin, Devrim Yilmaz, Matheus Grasselli, and Timothy J. Garrett. 2021. "Economists' Erroneous Estimates of Damages from Climate Change." *arXiv.org*.

Kelly, Erin, and John Rawls, eds. 2001. *Justice as Fairness: A Restatement*. Cambridge, Mass.: The Belknap Press of Harvard Univ. Press. *65034*.

Kelly, Thomas. 2023. "Evidence." In *The Stanford Encyclopedia of Philosophy*, edited by Edward N. Zalta. Metaphysics Research Lab, Stanford University. https://plato.stanford.edu/entries/evidence/.

Keppo, I., I. Butnar, N. Bauer, M. Caspani, O. Edelenbosch, J. Emmerling, P. Fragkos, et al. 2021. "Exploring the Possibility Space: Taking Stock of the Diverse Capabilities and Gaps in Integrated Assessment Models." *Environmental Research Letters* 16 (5): 053006. https://doi.org/10.1088/1748-9326/abe5d8.

Keyßer, Lorenz T., and Manfred Lenzen. 2021. "1.5 °c Degrowth Scenarios Suggest the Need for New Mitigation Pathways." *Nature Communications* 12 (1): 1–16. https://doi.org/10.1038/s41467-021-22884-9.

Kikstra, Jarmo S., Alessio Mastrucci, Jihoon Min, Keywan Riahi, and Narasimha D. Rao. 2021. "Decent Living Gaps and Energy Needs Around the World." *Environmental Research Letters* 16 (9): 095006. https://doi.org/10.1088/1748-9326/ac1c27.

Kirchin, Simon, ed. 2013. *Thick Concepts*. 1. ed. Mind Association Occasional Series. Oxford: Oxford University Press.

Kitcher, Philip. 2001. *Science, Truth, and Democracy*. Oxford Studies in Philosophy of Science. Oxford; New York: Oxford University Press.

Kitcher, Philip. 2011. *Science in a Democratic Society*. Amherst, New York: Prometheus Books.

Knopf, Brigitte, Gunnar Luderer, and Ottmar Edenhofer. 2011. "Exploring the Feasibility of Low Stabilization Targets." *Wiley Interdisciplinary Reviews: Climate Change* 2 (4): 617–26. https://doi.org/10.1002/wcc.124.

Köberle, Alexandre C. 2019. "The Value of BECCS in IAMs: A Review." *Current Sustainable/Renewable Energy Reports* 6 (4): 107–15. https://doi.org/10.1007/s40518-019-00142-3.

Kolodny, Niko. 2014a. "Rule over None i: What Justifies Democracy?" *Philosophy & Public Affairs* 42 (3): 195–229. https://doi.org/10.1111/papa.12035.

Kolodny, Niko. 2014b. "Rule over None II: Social Equality and the Justification of Democracy." *Philosophy & Public Affairs* 42 (4): 287–336. https://doi.org/10.1111/papa.12037.

Kolstad, Charles, Kevin Urama, John Broome, Annegrete Bruvoll, Micheline Cariño Olvera, Don Fullerton, Christian Gollier, et al. 2014. "Social, Economic, and Ethical Concepts and Methods." In *Climate Change 2014: Mitigation of Climate Change*, edited by Edenhofer, O., R. Pichs-Madruga, Y. Sokona, E. Farahani, S. Kadner, K. Seyboth, A. Adler, I. Baum, S. Brunner, P. Eickemeier, B. Kriemann, J. Savolainen, S. Schlömer, C. von Stechow, T. Zwickel and J.C. Minx, 211–82. Cambridge, UK & New York, NY, USA: Cambridge University Press.

Korsgaard, Christine M. 2018. *Fellow Creatures: Our Obligations to the Other Animals*. First edition. Uehiro Series in Practical Ethics. Oxford: Oxford University Press.

Kourany, Janet A., and Martin Carrier. 2020. "Introducing the Issues." In *Science and the Production of Ignorance*, edited by Janet A. Kourany and Martin Carrier, 3–26. Cambridge, Massachusetts; London,England: The MIT Press.

Kowarsch, Martin. 2016. *A Pragmatist Orientation for the Social Sciences in Climate Policy*. Vol. 323. Cham: Springer International Publishing. https://doi.org/10.1007/978-3-319-43281-6.

Kowarsch, Martin, and Ottmar Edenhofer. 2016. "Principles or Pathways? Improving the Contribution of Philosophical Ethics to Climate Policy." In *Climate Justice in a Non-Ideal World*, edited by Clare Heyward and Dominic Roser, 296–318. Oxford University Press. https://doi.org/10.1093/acprof:oso/9780198744047.003.0015.

Kowarsch, Martin, Christian Flachsland, Jennifer Garard, Jason Jabbour, and Pauline Riousset. 2017. "The Treatment of Divergent Viewpoints in Global Environmental Assessments." *Environmental Science & Policy* 77: 225–34. https://doi.org/10.1016/j.envsci.2017.04.001.

Kowarsch, Martin, Jennifer Garard, Pauline Riousset, Dominic Lenzi, Marcel J. Dorsch, Brigitte Knopf, Jan-Albrecht Harrs, and Ottmar Edenhofer. 2016. "Scientific Assessments to Facilitate Deliberative Policy Learning." *Palgrave Communications* 2 (1). https://doi.org/10.1057/palcomms.2016.92.

Kowarsch, Martin, Jason Jabbour, Christian Flachsland, Marcel T. J. Kok, Robert Watson, Peter M. Haas, Jan C. Minx, et al. 2017. "A Road Map for Global Environmental Assessments." *Nature Climate Change* 7 (6): 379–82. https://doi.org/10.1038/nclimate3307.

Kratzer, Angelika. 1977. "What 'Must' and 'Can' Must and Can Mean." *Linguistics and Philosophy* 1 (3): 337–55. https://doi.org/10.1007/bf00353453.

Krey, Volker, Fei Guo, Peter Kolp, Wenji Zhou, Roberto Schaeffer, Aayushi Awasthy, Christoph Bertram, et al. 2019. "Looking Under the Hood: A Comparison of Techno-Economic Assumptions Across National and Global Integrated Assessment Models." *Energy* 172: 1254–67. https://doi.org/10.1016/j.energy.2018.12.131.

Krey, Volker, P. Havlik, P. N. Kishimoto, O. Fricko, J. Zilliacus, M. Gidden, M. Strubegger, et al. 2020. "MESSAGEix-GLOBIOM Documentation – 2020 Release. Laxenburg, Austria." Technical Report. International Institute for Applied Systems Analysis. https://doi.org/10.22022/iacc/03-2021.17115.

Kuhn, Thomas. 1977. "Objectivity, Value Judgment, and Theory Choice." In *The Essential Tension: Selected Studies in Scientific Tradition and Change*, edited by Thomas Kuhn, 320–39. University of Chicago Press.

Kusch, Martin. 2020. *Relativism in the Philosophy of Science*. Cambridge University Press. https://doi.org/10.1017/9781108979504.

Lacey, Hugh. 1999. *Is Science Value Free?: Values and Scientific Understanding*. London; New York: Routledge.

Lacey, Hugh. 2005. *Values and Objectivity in Science: The Current Controversy about Transgenic Crops*. Lanham, MD: Lexington Books.

Lacey, Hugh. 2013. "Rehabilitating Neutrality." *Philosophical Studies* 163 (1): 77–83. https://doi.org/10.1007/s11098-012-0074-6.

Lamb, William F., and Julia K. Steinberger. 2017. "Human Well-Being and Climate Change Mitigation." *Wiley Interdisciplinary Reviews: Climate Change* 8 (6): e485. https://doi.org/10.1002/wcc.485.

Lashof, D. A., and D. A. Tirpak. 1990. "Policy Options for Stabilizing Global Climate." Edited by Hemisphere Publishing. New York.

Lawford-Smith, Holly. 2012. "The Feasibility of Collectives' Actions." *Australasian Journal of Philosophy* 90 (3): 453–67. https://doi.org/10.1080/00048402.2011.594446.

Lawford-Smith, Holly. 2013. "Understanding Political Feasibility." *Journal of Political Philosophy* 21 (3): 243–59. https://doi.org/10.1111/j.1467-9760.2012.00422.x.

Le Quéré, Corinne, Jan Ivar Korsbakken, Charlie Wilson, Jale Tosun, Robbie Andrew, Robert J. Andres, Josep G. Canadell, Andrew Jordan, Glen P. Peters, and Detlef P. van Vuuren. 2019. "Drivers of Declining CO2 Emissions in 18 Developed Economies." *Nature Climate Change* 9 (3): 213–17. https://doi.org/10.1038/s41558-019-0419-7.

Le Quéré, Corinne, and Asher Minns. 2016. "Where Next for Global Environmental Research? The Answer Is Future Earth." *Annales Des Mines—Responsabilité Et Environnement* N° 83 (3): 72–77. https://doi.org/10.3917/re1.083.0072.

Lee, Hoesung. 2015. "Turning the Focus to Solutions." *Science (New York, N.Y.)* 350 (6264): 1007. https://doi.org/10.1126/science.aad8954.

Leimbach, Marian, Anselm Schultes, Lavinia Baumstark, Anastasis Giannousakis, and Gunnar Luderer. 2017. "Solution Algorithms for Regional Interactions in Large-Scale Integrated Assessment Models of Climate Change." *Annals of Operations Research* 255 (1–2): 29–45. https://doi.org/10.1007/s10479-016-2340-z.

Lenzi, Dominic. 2018. "The Ethics of Negative Emissions." *Global Sustainability* 1 (E7): 1–8. https://doi.org/10.1017/SUS.2018.5.

Lenzi, Dominic. 2019. "Deliberating about Climate Change: The Case for 'Thinking and Nudging'." *Moral Philosophy and Politics* 6 (2): 313–36. https://doi.org/10.1515/mopp-2018-0034.

Lenzi, Dominic. 2021. "On the Permissibility (or Otherwise) of Negative Emissions." *Ethics, Policy & Environment* 24 (2): 1–14. https://doi.org/10.1080/21550085.2021.1885249.

Lenzi, Dominic, and Martin Kowarsch. 2021. "Integrating Justice in Climate Policy Assessments: Towards a Deliberative Transformation of Feasibility." In *Principles of Justice and Real-World Climate Politics*, edited by Sarah Kenehan and Corey Katz, 15–33. Rowman & Littlefield Publishers.

Lenzi, Dominic, William F. Lamb, Jérôme Hilaire, Martin Kowarsch, and Jan C. Minx. 2018. "Don't Deploy Negative Emissions Technologies Without Ethical Analysis." *Nature* 561 (7723): 303–5. https://doi.org/10.1038/d41586-018-06695-5.

Leuschner, Anna. 2012. "Pluralism and Objectivity: Exposing and Breaking a Circle." *Studies in History and Philosophy of Science Part A* 43 (1): 191–98. https://doi.org/10.1016/j.shpsa.2011.12.030.

Levi, Isaac. 1960. "Must the Scientist Make Value Judgments?" *The Journal of Philosophy* 57 (11): 345. https://doi.org/10.2307/2023504.

Livingston, Jasmine E., and Markku Rummukainen. 2020. "Taking Science by Surprise: The Knowledge Politics of the IPCC Special Report on 1.5 Degrees." *Environmental Science & Policy* 112: 10–16. https://doi.org/10.1016/j.envsci.2020.05.020.

Lockhart, Ted. 2000. *Moral Uncertainty and Its Consequences*. New York NY u.a.: Oxford Univ. Press.

Loftus, Peter J., Armond M. Cohen, Jane C. S. Long, and Jesse D. Jenkins. 2015. "A Critical Review of Global Decarbonization Scenarios: What Do They Tell Us about Feasibility?" *Wiley Interdisciplinary Reviews: Climate Change* 6 (1): 93–112. https://doi.org/10.1002/wcc.324.

Longino, Helen E. 1979. "Evidence and Hypothesis: An Analysis of Evidential Relations." *Philosophy of Science* 46 (1): 35–56. https://doi.org/10.1086/288849.

Longino, Helen E. 1990. *Science as Social Knowledge: Values and Objectivity in Scientific Inquiry*. Princeton, N.J.: Princeton Univ. Press.

Longino, Helen E. 1995. "Gender, Politics, and the Theoretical Virtues." *Synthese* 104 (3): 383–97. https://doi.org/10.1007/bf01064506.

Low, Sean, and Stefan Schäfer. 2020. "Is Bio-Energy Carbon Capture and Storage (BECCS) Feasible? The Contested Authority of Integrated Assessment Modeling." *Energy Research & Social Science* 60: 101326. https://doi.org/10.1016/j.erss.2019.101326.

Luckner, Andreas. 2005. *Klugheit*. de Gruyter. https://doi.org/10.1515/9783110898309.

Luderer, Gunnar, Nico Bauer, Lavinia Baumstark, Christoph Bertram, Marian Leimbach, Robert Pietzcker, Jessica Strefler, et al. 2020. "REMIND—REgional Model of INvestments and Development: Repository Code: Https://Github.com/Remindmodel/Remind." PIK. https://www.pik-potsdam.de/research/transformation-pathways/models/remind.

Luderer, Gunnar, Valentina Bosetti, Jan Steckel, Henri Waisman, N. Bauer, E. de Cian, Marian Leimbach, O. Sassi, and Massimo Tavoni. 2009. "The Economics of Decarbonzation: Results from the RECIPE Model Intercomparison. RECIPE Background Paper." RECIPE Background Paper. Potsdam: Potsdam-Institut für Klimafolgenforschung.

Luderer, Gunnar, Marian Leimbach, Nico Bauer, Elmar Kriegler, Lavinia Baumstark, Christoph Bertram, Anastasis Giannousakis, et al. 2015. "Description of the REMIND Model (Version 1.6)." Potsdam.

Luderer, Gunnar, Robert C. Pietzcker, Christoph Bertram, Elmar Kriegler, Malte Meinshausen, and Ottmar Edenhofer. 2013. "Economic Mitigation Challenges: How Further Delay Closes the Door for Achieving Climate Targets." *Environmental Research Letters* 8 (3): 034033. https://doi.org/10.1088/1748-9326/8/3/034033.

Lumer, Christoph. 2005. "Prioritarian Welfare Functions – an Elaboration and Justification." *Working Paper University of Siena Department of Philosophy and Social Sciences.*

MacAskill, Michael. 2020. *Moral Uncertainty.* Oxford: Oxford University Press.

Majone, Giandomenico. 1974. "The Role of Constraints in Policy Analysis." *Quality and Quantity* 8 (1). https://doi.org/10.1007/BF00205865.

Majone, Giandomenico. 1975. "On the Notion of Political Feasibility." *European Journal of Political Research* 3 (3): 259–74. https://doi.org/10.1111/j.1475-6765.1975.tb00780.x.

Mandelkern, Matthew, Ginger Schultheis, and David Boylan. 2017. "Agentive Modals." *The Philosophical Review* 126 (3): 301–43. https://doi.org/10.1215/00318108-3878483.

Mansbridge, Jane. 2003. "Rethinking Representation." *The American Political Science Review* 97 (4): 515–28.

Marcucci, Adriana, Socrates Kypreos, and Evangelos Panos. 2017. "The Road to Achieving the Long-Term Paris Targets: Energy Transition and the Role of Direct Air Capture." *Climatic Change* 144 (2): 181–93. https://doi.org/10.1007/s10584-017-2051-8.

Masson-Delmotte, V., P. Zhai, H.-O. Pörtner, D. Roberts, J. Skea, P.R. Shukla, et al., eds. 2018. "Global Warming of 1.5°c. An IPCC Special Report on the Impacts of Global Warming of 1.5°c Above Pre-Industrial Levels and Related Global Greenhouse Gas Emission Pathways, in the Context of Strengthening the Global Response to the Threat of Climate Change, Sustainable Development, and Efforts to Eradicate Poverty." Cambridge, UK; New York, NY, USA: IPCC; Cambridge University Press.

McCollum, David L., Wenji Zhou, Christoph Bertram, Harmen-Sytze de Boer, Valentina Bosetti, Sebastian Busch, Jacques Després, et al. 2018. "Energy Investment Needs for Fulfilling the Paris Agreement and Achieving the Sustainable Development Goals." *Nature Energy* 3 (7): 589–99. https://doi.org/10.1038/s41560-018-0179-z.

McLaren, Duncan, and Nils Markusson. 2020. "The Co-Evolution of Technological Promises, Modelling, Policies and Climate Change Targets." *Nature Climate Change* 10 (5): 392–97. https://doi.org/10.1038/s41558-020-0740-1.

McShane, Katie. 2016. "Anthropocentrism in Climate Ethics and Policy." *Midwest Studies In Philosophy* 40 (1): 189–204. https://doi.org/10.1111/misp.12055.

McShane, Katie. 2018. "Why Animal Welfare Is Not Biodiversity, Ecosystem Services, or Human Welfare: Toward a More Complete Assessment of Climate Impacts." *Les Ateliers de l'Éthique / the Ethics Forum* 13 (1): 43–64. https://doi.org/10.7202/1055117ar.

McTernan, Emily. 2019. "Justice, Feasibility, and Social Science as It Is." *Ethical Theory and Moral Practice* 22 (1): 27–40. https://doi.org/10.1007/s10677-018-9970-y.

Meadows, Donella H., Dennis Meadows, Jorgen Randers, and William W. Behrens. 1972. *The Limits to Growth: A Report for the Club of Rome's Project on the Predicament of Mankind*. New York, NY: Club of Rome; Universe Books.

Merriam Webster. 2023. "Feasible." https://www.merriam-webster.com/dictionary/feasible.

Mill, John Stuart. 1861. *Utilitarianism*. Collected Works of John Stuart Mill, in 33 Vols. Toronto: The University of Toronto Press.

Miller, Alexander. 2013. *Contemporary Metaethics: An Introduction*. 2. edition. Cambridge; Malden, MA: Polity Press.

Miller, David. 2013. *Justice for Earthlings: Essays in Political Philosophy*. Cambridge: Cambridge University Press. https://doi.org/10.1017/CBO9781139236898.

Mills, Charles W. 2005. "'Ideal Theory' as Ideology." *Hypatia* 20 (3): 165–83. https://doi.org/10.1111/j.1527-2001.2005.tb00493.x.

Millward-Hopkins, Joel, and Yannick Oswald. 2023. "Reducing Global Inequality to Secure Human Wellbeing and Climate Safety: A Modelling Study." *The Lancet. Planetary Health* 7 (2): e147–54. https://doi.org/10.1016/S2542-5196(23)00004-9.

Mintz-Woo, Kian. 2018a. "Moral Uncertainty over Policy Evaluation." Dissertation, Graz: Karl-Franzens-Universität Graz. https://unipub.uni-graz.at/obvugrhs/2581643.

Mintz-Woo, Kian. 2018b. "Moral Uncertainty over Policy Evaluation." *Erasmus Journal for Philosophy and Economics* 11 (2): 291–94. https://doi.org/10.23941/ejpe.v11i2.351.

Mintz-Woo, Kian. 2021a. "A Philosopher's Guide to Discounting." In *Philosophy and Climate Change*, edited by Mark Budolfson, Tristram McPherson, and David Plunkett, 90–110. Oxford University Press. https://doi.org/10.1093/oso/9780198796282.003.0005.

Mintz-Woo, Kian. 2021b. "The Ethics of Measuring Climate Change Impacts." In *The Impacts of Climate Change*, 521–35. Elsevier. https://doi.org/10.1016/B978-0-12-822373-4.00023-9.

Minx, Jan C., William F. Lamb, Max W. Callaghan, Sabine Fuss, Jérôme Hilaire, Felix Creutzig, Thorben Amann, et al. 2018. "Negative Emissions—Part 1: Research Landscape and Synthesis." *Environmental Research Letters* 13 (6): 063001. https://doi.org/10.1088/1748-9326/aabf9b.

Moellendorf, Darrel. 2011. "Why Global Inequality Matters." *Journal of Social Philosophy* 42 (1): 99–109.

Moellendorf, Darrel. 2013. "Discounting the Future and the Morality in Climate Change Economics." In *The Moral Challenge of Dangerous Climate Change*, edited by Darrel Moellendorf, 90–122. New York: Cambridge University Press. https://doi.org/10.1017/CBO9781139083652.005.

Mogensen, Andreas L. 2022. "The Only Ethical Argument for Positive δ? Partiality and Pure Time Preference." *Philosophical Studies* 179 (9): 2731–50. https://doi.org/10.1007/s11098-022-01792-8.

Möller, Niklas. 2012. "The Concepts of Risk and Safety." In *Handbook of Risk Theory*, edited by Sabine Roeser, 55–85. Springer Reference. Dordrecht: Springer. https://doi.org/10.1007/978-94-007-1433-5_3.

Moore, G. E. 1903 [1993]. *Principia Ethica*. Rev. ed. /edited and with an introduction by Thomas Baldwin. Cambridge: Cambridge University Press.

Nagel, Ernest. 1961. *The Structure of Science: Problems in the Logic of Scientific Explanation*. New York, Chicago, San Fransisco Atlanta: Harcourt, Brace & World, Inc.

Neef-Max, Manfred. 1991. *Human Scale Development: Conception, Application and Further Reflections.* New York: Apex press.

New York Times. 2023. "Search Function." New York. https://www.nytimes.com/.

Nielsen, Kristian S., Paul C. Stern, Thomas Dietz, Jonathan M. Gilligan, Detlef P. van Vuuren, Maria J. Figueroa, Carl Folke, et al. 2020. "Improving Climate Change Mitigation Analysis: A Framework for Examining Feasibility." *One Earth* 3 (3): 325–36. https://doi.org/10.1016/j.oneear.2020.08.007.

NOAA. 2024. "Climate Change: Atmospheric Carbon Dioxide." https://www.climate.gov/news-features/understanding-climate/climate-change-atmospheric-carbon-dioxide.

Nordhaus, William D. 1979. *The Efficient Use of Energy Resources.* Vol. 26. Monograph / Cowles Foundation for Research in Economics. New Haven: Yale Univ. Press.

Nordhaus, William D. 1997. "Discounting in Economics and Climate Change; an Editorial Comment." *Climatic Change* 37 (2): 315–28. https://doi.org/10.1023/A:1005347001731.

Nordhaus, William D. 2007. "A Review of the Stern Review on the Economics of Climate Change." *Journal of Economic Literature* 45 (3): 686–702.

Nordhaus, William D. 2008. *A Question of Balance: Weighing the Options on Global Warming Policies.* New Haven; London: Yale University Press.

Nordhaus, William D. 2010. "Economic Aspects of Global Warming in a Post-Copenhagen Environment." *Proceedings of the National Academy of Sciences of the United States of America* 107 (26): 11721–26. https://doi.org/10.1073/pnas.1005985107.

Nordhaus, William D. 2013. "Integrated Economic and Climate Modeling." In *Handbook of Computable General Equilibrium Modeling SET, Vols. 1A and 1B*, edited by Peter B. Dixon and Dale W. Jorgenson, 1:1069–1131. Handbook of Computable General Equilibrium Modeling. Elsevier. https://doi.org/10.1016/B978-0-444-59568-3.00016-X.

Nordhaus, William D. 2017. "Revisiting the Social Cost of Carbon." *Proceedings of the National Academy of Sciences of the United States of America* 114 (7): 1518–23. https://doi.org/10.1073/pnas.1609244114.

Nordhaus, William D. 2019. "Biographical." Edited by Watson Publishing. International LLC, Sagamore Beach. https://www.nobelprize.org/prizes/economic-sciences/2018/nordhaus/biographical/.

Nussbaum, Martha C. 2012. *Women and Human Development.* Cambridge University Press. https://doi.org/10.1017/CBO9780511841286.

Nussbaum, Martha C. 2022. *Justice for Animals: Our Collective Responsibility.* First Simon & Schuster hardcover edition. New York; London; Toronto: Simon & Schuster.

O'Neill, Brian C., Elmar Kriegler, Keywan Riahi, Kristie L. Ebi, Stephane Hallegatte, Timothy R. Carter, Ritu Mathur, and Detlef P. van Vuuren. 2014. "A New Scenario Framework for Climate Change Research: The Concept of Shared Socioeconomic Pathways: Climatic Change, 122(3), 387–400." *Climatic Change* 122 (3): 387–400. https://doi.org/10.1007/S10584-013-0905-2.

O'Neill, Daniel W., Andrew L. Fanning, William F. Lamb, and Julia K. Steinberger. 2018. "A Good Life for All Within Planetary Boundaries." *Nature Sustainability* 1 (2): 88–95. https://doi.org/10.1038/s41893-018-0021-4.

O'Neill, Onora. 1987. "Abstraction, Idealization and Ideology in Ethics." *Royal Institute of Philosophy Lecture Series* 22: 55–69. https://doi.org/10.1017/S0957042X00003667.

O'Neill, Onora. 1996. *Towards Justice and Virtue: A Constructive Account of Practical Reasoning.* Cambridge: Cambridge University Press.

Oei, Pao-Yu, Catharina Rieve, Philipp Herpich, and Claudia Kemfert. 2023. "1,5-Grad-Grenze von Paris Nach lützerath: CO 2 -Budget für Den Tagebau Garzweiler II Studie Der FossilExit Forschungsgruppe." FossilExit Forschungsgruppe.

Oreskes, Naomi. 1998. "Evaluation (Not Validation) of Quantitative Models." *Environmental Health Perspectives* 106 Suppl 6: 1453–60. https://doi.org/10.1289/ehp.98106s61453.

Oreskes, Naomi, and Erik M. Conway. 2010. *Merchants of Doubt: How a Handful of Scientists Obscured the Truth on Issues from Tobacco Smoking to Global Warming.* Paperback. ed. London: Bloomsbury.

Pachauri, Shonali, Setu Pelz, Christoph Bertram, Silvie Kreibiehl, Narasimha D. Rao, Youba Sokona, and Keywan Riahi. 2022. "Fairness Considerations in Global Mitigation Investments." *Science (New York, N.Y.)* 378 (6624): 1057–59. https://doi.org/10.1126/science.adf0067.

Page, Edward A. 2008. "Distributing the Burdens of Climate Change." *Environmental Politics* 17 (4): 556–75. https://doi.org/10.1080/09644010802193419.

Parfit, Derek. 1984. *Reasons and Persons.* Oxford: Oxford University Press.

Parkinson, John. 2003. "Legitimacy Problems in Deliberative Democracy." *Political Studies* 51 (1): 180–96. https://doi.org/10.1111/1467-9248.00419.

Patterson, James J., Thomas Thaler, Matthew Hoffmann, Sara Hughes, Angela Oels, Eric Chu, Aysem Mert, Dave Huitema, Sarah Burch, and Andy Jordan. 2018. "Political Feasibility of 1.5°c Societal Transformations: The Role of Social Justice." *Current Opinion in Environmental Sustainability* 31: 1–9. https://doi.org/10.1016/j.cosust.2017.11.002.

Pedersen, Jiesper Strandsbjerg Tristan, Filipe Duarte Santos, Detlef van Vuuren, Joyeeta Gupta, Ricardo Encarnação Coelho, Bruno A. Aparício, and Rob Swart. 2021. "An Assessment of the Performance of Scenarios Against Historical Global Emissions for IPCC Reports." *Global Environmental Change* 66: 102199. https://doi.org/10.1016/j.gloenvcha.2020.102199.

Peter, Fabienne. 2023. "Political Legitimacy: 2017." In *The Stanford Encyclopedia of Philosophy,* edited by Edward N. Zalta. Metaphysics Research Lab, Stanford University. https://plato.stanford.edu/entries/legitimacy/#PolLegDem.

Peters, Glen P. 2016. "The 'Best Available Science' to Inform 1.5 °c Policy Choices." *Nature Climate Change* 6 (7): 646–49. https://doi.org/10.1038/nclimate3000.

Pfenninger, Stefan, Lion Hirth, Ingmar Schlecht, Eva Schmid, Frauke Wiese, Tom Brown, Chris Davis, et al. 2018. "Opening the Black Box of Energy Modelling: Strategies and Lessons Learned." *Energy Strategy Reviews* 19: 63–71. https://doi.org/10.1016/j.esr.2017.12.002.

Pidcock, Roz. 16.08.2016. "IPCC Special Report to Scrutinise 'Feasibility' of 1.5C Climate Goal." *CarbonBrief* 2016 (16.08.2016). https://www.carbonbrief.org/ipcc-special-report-feasibility-1point5/.

Pielke, Roger. 2018. "Opening up the Climate Policy." *Issues in Science and Technology* 34 (4): 30–36.

PIK. 2022. "The Senses Toolkit." https://climatescenarios.org/toolkit/.

Piketty, Thomas. 2014. *Capital in the Twenty-First Century.* Cambridge, Massachusetts; London: The Belknap Press of Harvard University Press.

Piketty, Thomas. 2015. *The Economics of Inequality.* Cambridge, Massachusetts; London, England: The Belknap Press of Harvard University Press.

Pindyck, Robert. 2013. "Climate Change Policy: What Do the Models Tell Us?" *NBER Working Paper*, no. 19244. https://doi.org/10.3386/w19244.

Pitkin, Hanna Fenichel. 1972. *The Concept of Representation.* Berkeley: University of California Press.

Posner, Eric A., and David Weisbach. 2010. *Climate Change Justice.* Princeton, N.J.: Princeton University Press.

Proctor, Robert N., and Londa Schiebinger, eds. 2008. *Agnotology: The Making and Unmaking of Ignorance.* Stanford, California: Stanford University Press.

Purvis, Ben. 2021. "Modelling Global Futures: A Comparison of 'Limits to Growth' and the Use of Integrated Assessment Models Within the Climate Literature." *2021 Conference of the System Dynamics Society.*

Putnam, Hilary. 2002. *The Collapse of the Fact/Value Dichotomy and Other Essays.* Cambridge, Mass.; London, England: Harvard University Press.

Räikkä, Juha. 1998. "The Feasibility Condition in Political Theory." *Journal of Political Philosophy* 6 (1): 27–40. https://doi.org/10.1111/1467-9760.00044.

Ramsey, F. P. 1928. "A Mathematical Theory of Saving." *The Economic Journal* 38 (152): 543. https://doi.org/10.2307/2224098.

Rao, Narasimha D., and Jihoon Min. 2018. "Decent Living Standards: Material Prerequisites for Human Wellbeing." *Social Indicators Research* 138 (1): 225–44. https://doi.org/10.1007/s11205-017-1650-0.

Rao, Narasimha D., Bas J. van Ruijven, Keywan Riahi, and Valentina Bosetti. 2017. "Improving Poverty and Inequality Modelling in Climate Research." *Nature Climate Change* 7 (12): 857–62. https://doi.org/10.1038/s41558-017-0004-x.

Rawls, John. 1999. *A Theory of Justice.* Rev. ed. Cambridge, MA: Harvard University Press.

Realmonte, Giulia, Laurent Drouet, Ajay Gambhir, James Glynn, Adam Hawkes, Alexandre C. Köberle, and Massimo Tavoni. 2019. "An Inter-Model Assessment of the Role of Direct Air Capture in Deep Mitigation Pathways." *Nature Communications* 10 (1): 3277. https://doi.org/10.1038/s41467-019-10842-5.

Regan, Tom. 2004 [1983]. *The Case for Animal Rights.* Berkeley, Los Angeles: University of California Press.

Reiss, Julian. 2013. *Philosophy of Economics: A Contemporary Introduction.* 1st ed. Routledge Contemporary Introductions to Philosophy. New York: Routledge.

Reiss, Julian. 2017. "Fact-Value Entanglement in Positive Economics." *Journal of Economic Methodology* 24 (2): 134–49. https://doi.org/10.1080/1350178X.2017.1309749.

Research Council. 1983. *Changing Climate.* Washington, D.C.: National Academies Press. https://doi.org/10.17226/18714.

Resnik, David B. 2000. "Financial Interests and Research Bias." *Perspectives on Science* 8 (3): 255–85. https://doi.org/10.1162/106361400750340497.

RFF-CMCC-EIEE. 2023. "The WITCH Model." Milan: RFF-CMCC-EIEE European Institute on Economics and the Environment. https://www.witchmodel.org/.

Riahi, Keywan, Christoph Bertram, Daniel Huppmann, Joeri Rogelj, Valentina Bosetti, Anique-Marie Cabardos, Andre Deppermann, et al. 2021. "Cost and Attainability of Meeting Stringent Climate Targets Without Overshoot." *Nature Climate Change* 11 (12): 1063–69. https://doi.org/10.1038/s41558-021-01215-2.

Riahi, Keywan, Elmar Kriegler, Nils Johnson, Christoph Bertram, Michel den Elzen, Jiyong Eom, Michiel Schaeffer, et al. 2015. "Locked into Copenhagen Pledges — Implications of Short-Term Emission Targets for the Cost and Feasibility of Long-Term Climate Goals." *Technological Forecasting and Social Change* 90: 8–23. https://doi.org/10.1016/j.techfore.2013.09.016.

Riahi, Keywan, R. Schaeffer, J. Arango, K. Calvin, C. Guivarch, T. Hasegawa, K. Jiang, et al. 2022. "2022: Mitigation Pathways Compatible with Long-Term Goals." In *IPCC, 2022: Climate Change 2022: Mitigation of Climate Change. Contribution of Working Group III to the Sixth Assessment Report of the Intergovernmental Panel on Climate Change.* Cambridge, UK; New York, NY, USA: Cambridge University Press. https://doi.org/10.1017/9781009157926.005.

Riahi, Keywan, Detlef P. van Vuuren, Elmar Kriegler, Jae Edmonds, Brian C. O'Neill, Shinichiro Fujimori, Nico Bauer, et al. 2017. "The Shared Socioeconomic Pathways and Their Energy, Land Use, and Greenhouse Gas Emissions Implications: An Overview." *Global Environmental Change* 42: 153–68. https://doi.org/10.1016/j.gloenvcha.2016.05.009.

Rieve, Catharina, Philipp Herpich, Luna Brandes, Pao-Yu Oei, Claudia Kemfert, and Christian R. von Hirschhausen. 2021. *Kein Grad Weiter—Anpassung Der Tagebauplanung Im Rheinischen Braunkohlerevier Zur Einhaltung Der 1,5-Grad-Grenze: Im Auftrag von Alle dörfer Bleiben (Kib e.v.).* Vol. 169. DIW Berlin. Berlin: DIW Berlin Deutsches Institut für Wirtschaftsforschung. https://www.diw.de/documents/publikationen/73/diw_01.c.819609.de/diwkompakt_2021-169.pdf.

Roberts, M. A. 2023. "The Nonidentity Problem." In *The Stanford Encyclopedia of Philosophy*, edited by Edward N. Zalta. Metaphysics Research Lab, Stanford University.

Robertson, Simon. 2021. "Transparency, Trust, and Integrated Assessment Models: An Ethical Consideration for the Intergovernmental Panel on Climate Change." *Wiley Interdisciplinary Reviews: Climate Change* 12 (1). https://doi.org/10.1002/wcc.679.

Robertson, Simon. 2022. "AN OPEN LETTER: Re. An Assay into Scientific Integrity of the IPCC and IAMs in the AR6—Matters of Legitimate Public Interest." https://www.researchgate.net/publication/364289578_AN_OPEN_LETTER_Re_An_assay_into_scientific_integrity_of_the_IPCC_and_IAMs_in_the_AR6_-_matters_of_legitimate_public_interest.

Roelfsema, Mark, Heleen L. van Soest, Michel den Elzen, Heleen de Coninck, Takeshi Kuramochi, Mathijs Harmsen, Ioannis Dafnomilis, Niklas Höhne, and Detlef P. van Vuuren. 2022. "Developing Scenarios in the Context of the Paris Agreement and Application in the Integrated Assessment Model IMAGE: A Framework for Bridging the Policy-Modelling Divide." *Environmental Science & Policy* 135: 104–16. https://doi.org/10.1016/j.envsci.2022.05.001.

Roemer, John E. 2011. "The Ethics of Intertemporal Distribution in a Warming Planet." *Environmental and Resource Economics* 48 (3): 363–90. https://doi.org/10.1007/s10640-010-9414-1.

Rogelj, Joeri, Daniel Huppmann, Volker Krey, Keywan Riahi, Leon Clarke, Matthew Gidden, Zebedee Nicholls, and Malte Meinshausen. 2019. "A New Scenario Logic for the Paris Agreement Long-Term Temperature Goal." *Nature* 573 (7774): 357–63. https://doi.org/10.1038/s41586-019-1541-4.

Rogelj, Joeri, Alexander Popp, Katherine V. Calvin, Gunnar Luderer, Johannes Emmerling, David Gernaat, Shinichiro Fujimori, et al. 2018. "Scenarios Towards Limiting Global Mean Temperature Increase Below 1.5 °c." *Nature Climate Change* 8 (4): 325–32. https://doi.org/10.1038/s41558-018-0091-3.

Rogelj, Joeri, D. Shindell, K. Jiang, S. Fifita, P. Forster, V. Ginzburg, C. Handa, et al. 2018. "Mitigation Pathways Compatible with 1.5°c in the Context of Sustainable Development." In *Global Warming of 1.5°c*, edited by Masson-Delmotte, V., P. Zhai, H.-O. Pörtner, D. Roberts, J. Skea, P.R. Shukla, et al., 93–174. Cambridge, UK; New York, NY, USA: Cambridge University Press. https://doi.org/10.1017/9781009157940.004.

Rosen, Richard A. 2015. "IAMs and Peer Review." *Nature Climate Change* 5 (5): 390. https://doi.org/10.1038/nclimate2582.

Rosen, Richard A., and Edeltraud Guenther. 2015. "The Economics of Mitigating Climate Change: What Can We Know?" *Technological Forecasting and Social Change* 91: 93–106. https://doi.org/10.1016/j.techfore.2014.01.013.

Roser, Dominic. 2009. "The Discount Rate – a Small Number with a Big Impact." In *Applied Ethics: Life, Environment and Society*, edited by Center for Applied Ethics and Philosophy, 12–27. Sapporo: The Center for Applied Ethics and Philosophy, Hokkaido University.

Roser, Dominic. 2015. "Climate Justice in the Straitjacket of Feasibility." In *The Politics of Sustainability*, edited by Dieter Birnbacher and May Thorseth, 5–21. Abingdon, Oxon; New York, NY: Routledge.

Roser, Dominic. 2016. "Reducing Injustice Within the Bounds of Motivation." In *Climate Justice in a Non-Ideal World*, edited by Clare Heyward and Dominic Roser, 83–103. Oxford University Press. https://doi.org/10.1093/acprof:oso/9780198744047.003.0005.

Roser, Dominic, and Christian Seidel. 2015. *Ethik Des Klimawandels: Eine Einführung*. 2. Auflage. Darmstadt, Germany: Wissenschaftliche Buchgesellschaft.

Rotmans, Jan, and Marjolein van Asselt. 1996. "Integrated Assessment: A Growing Child on Its Way to Maturity." *Climatic Change* 34 (3–4): 327–36. https://doi.org/10.1007/BF00139296.

Rowland, Richard. 2019. *The Normative and the Evaluative*. Oxford University Press. https://doi.org/10.1093/oso/9780198833611.001.0001.

Rubiano Rivadeneira, Natalia, and Wim Carton. 2022. "(In)justice in Modelled Climate Futures: A Review of Integrated Assessment Modelling Critiques Through a Justice Lens." *Energy Research & Social Science* 92: 102781. https://doi.org/10.1016/j.erss.2022.102781.

Rudner, Richard. 1953. "The Scientist Qua Scientist Makes Value Judgments." *Philosophy of Science*, no. 20 (1): 1–6.

Saheb, Yamina, Kai Kuhnhenn, and Juliane Schumacher. 08.06.2022. "'It's a Very Western Vision of the World': Interview." https://www.rosalux.de/en/news/id/47045/its-a-very-western-vision-of-the-world.

Saward, Michael. 2009. "Authorisation and Authenticity: Representation and the Unelected*." *Journal of Political Philosophy* 17 (1): 1–22. https://doi.org/10.1111/j.1467-9760.2008.00309.x.

Scanlon, T. M. 1998. *What We Owe to Each Other*. Cambridge, Mass.: Harvard University Press.

Scanlon, Thomas. 2018. *Why Does Inequality Matter?* First edition. Uehiro Series in Practical Ethics. Oxford; New York: Oxford University Press.

Schellnhuber, Hans Joachim, Stefan Rahmstorf, and Ricarda Winkelmann. 2016. "Why the Right Climate Target Was Agreed in Paris." *Nature Climate Change* 6 (7): 649–53. https://doi.org/10.1038/nclimate3013.

Schienke, Erich W., Seth D. Baum, Nancy Tuana, Kenneth J. Davis, and Klaus Keller. 2011. "Intrinsic Ethics Regarding Integrated Assessment Models for Climate Management." *Science and Engineering Ethics* 17 (3): 503–23. https://doi.org/10.1007/s11948-010-9209-3.

Schuppert, Fabian. 2021. "Making the Great Climate Transition: Between Justice and Feasibility." In *Climate Justice and Feasibility: Moral and Practical Concerns in a Warming World*, edited by Sarah Kenehan and Corey Katz. Rowman & Littlefield Publisher.

Schüssler, Rudolf. 2011. "Climate Justice: A Question of Historic Responsibility?" *Journal of Global Ethics* 7 (3): 261–78. https://doi.org/10.1080/17449626.2011.635682.

Sen, Amartya. 1984. *Resources, Values and Development.* Oxford: Blackwell.

Sherwood, S. C., M. J. Webb, J. D. Annan, K. C. Armour, P. M. Forster, J. C. Hargreaves, G. Hegerl, et al. 2020. "An Assessment of Earth's Climate Sensitivity Using Multiple Lines of Evidence." *Reviews of Geophysics* 58 (4). https://doi.org/10.1029/2019RG000678.

Shockley, Kenneth. 2012. "Thinning the Thicket." *Environmental Ethics* 34 (3): 227–46. https://doi.org/10.5840/enviroethics201234320.

Shue, Henry. 1999. "Global Environment and International Inequality." *International Affairs (Royal Institute of International Affairs 1944-)* 75 (3): 531–45.

Shue, Henry. 2017. "Climate Dreaming: Negative Emissions, Risk Transfer, and Irreversibility." *Journal of Human Rights and the Environment* 8 (2): 203–16. https://doi.org/10.4337/jhre.2017.02.02.

Simmons, A. John. 2010. "Ideal and Nonideal Theory." *Philosophy & Public Affairs* 38 (1): 5–36. https://doi.org/10.1111/j.1088-4963.2009.01172.x.

Singer, Peter. 2009 [1990]. *Animal Liberation: The Definitive Classic of the Animal Movement.* New York, NY: Ecco Book/Harper Perennial.

Singer, Peter. 2010. "One Atmosphere." In *Climate Ethics: Essential Readings*, edited by Stephen Gardiner, Simon Caney, Dale Jamieson, and Henry Shue, 182–99. Oxford University Press.

Skea, Jim, Priyadarshi Shukla, Alaa Al Khourdajie, and David McCollum. 2021. "Intergovernmental Panel on Climate Change: Transparency and Integrated Assessment Modeling." *Wiley Interdisciplinary Reviews: Climate Change* 12 (5). https://doi.org/10.1002/wcc.727.

Smith, Noah. 2021. "Why Has Climate Economics Failed Us?" Noahpinion Blog. https://noahpinion.substack.com/p/why-has-climate-economics-failed?fbclid=IwAR273Zn6Wc lvfiDH2CC3IuVjmmUyDbODovYnHP6f5yce0C_mHCdx616VNaY.

Southwood, Nicholas. 2018. "The Feasibility Issue." *Philosophy Compass* 13 (8): e12509. https://doi.org/10.1111/phc3.12509.

Southwood, Nicholas. 2019. "Feasibility as a Constraint on 'Ought All-Things-Considered,' but Not on 'Ought as a Matter of Justice'?" *The Philosophical Quarterly* 69 (276): 598–616. https://doi.org/10.1093/pq/pqz012.

Southwood, Nicholas. 2022. "Feasibility as Deliberation–Worthiness." *Philosophy & Public Affairs* 50 (1): 121–62. https://doi.org/10.1111/papa.12206.

Southwood, Nicholas, and David Wiens. 2016. "'Actual' Does Not Imply 'Feasible'." *Philosophical Studies* 173 (11): 3037–60. https://doi.org/10.1007/s11098-016-0649-8.

SRU. 2020. "Umweltgutachten 2020: Für Eine Entschlossene Umweltpolitik in Deutschland Und Europa." Berlin: Sachverständigenrat für Umweltfragen.

Stanton, Elizabeth A. 2009. "Negishi Welfare Weights: The Mathematics of Global Inequality." SEI Working Paper. Somerville, MA.

Stanton, Elizabeth A., Frank Ackerman, and Sivan Kartha. 2009. "Inside the Integrated Assessment Models: Four Issues in Climate Economics." *Climate and Development* 1 (2): 166–84.

Steel, Daniel. 2010. "Epistemic Values and the Argument from Inductive Risk." *Philosophy of Science* 77 (1): 14–34. https://doi.org/10.1086/650206.

Stehfest, Elke, ed. 2014. *Integrated Assessment of Global Environmental Change with IMAGE 3.0: Model Description and Policy Applications.* The Hague: PBL Netherlands Environmental Assessment Agency.

Stehfest, Elke, Detlef van Vuuren, Lex Bouwman, and Tom Kram. 2021. "IMAGE 3.2 Documentation: IMAGE Integrated Model to Assess the Global Environment." PBL Netherlands Environmental. https://models.pbl.nl/image/index.php/.

Stemplowska, Zofia. 2016. "Feasibility: Individual and Collective." *Social Philosophy and Policy* 33 (1–2): 273–91. https://doi.org/10.1017/S0265052516000273.

Stemplowska, Zofia. 2020. "The Incentives Account of Feasibility." *Philosophical Studies.* https://doi.org/10.1007/s11098-020-01530-y.

Stern, Nicholas. 2007. *The Economics of Climate Change.* Cambridge: Cambridge University Press. https://doi.org/10.1017/CBO9780511817434.

Stern, Paul C., Thomas Dietz, Kristian S. Nielsen, Wei Peng, and Michael P. Vandenbergh. 2023. "Feasible Climate Mitigation." *Nature Climate Change* 13 (1): 6–8. https://doi.org/10.1038/s41558-022-01563-7.

Sterner, Thomas, and U. Martin Persson. 2008. "An Even Sterner Review: Introducing Relative Prices into the Discounting Debate." *Review of Environmental Economics and Policy* 2 (1): 61–76. https://doi.org/10.1093/reep/rem024.

Strefler, Jessica, Nico Bauer, Florian Humpenöder, David Klein, Alexander Popp, and Elmar Kriegler. 2021. "Carbon Dioxide Removal Technologies Are Not Born Equal." *Environmental Research Letters* 16 (7): 074021. https://doi.org/10.1088/1748-9326/ac0a11.

Strefler, Jessica, Nico Bauer, Elmar Kriegler, Alexander Popp, Anastasis Giannousakis, and Ottmar Edenhofer. 2018. "Between Scylla and Charybdis: Delayed Mitigation Narrows the Passage Between Large-Scale CDR and High Costs." *Environmental Research Letters* 13 (4): 044015. https://doi.org/10.1088/1748-9326/aab2ba.

Swift, Adam. 2008. "The Value of Philosophy in Nonideal Circumstances." *Social Theory and Practice* 34 (3): 363–87. https://doi.org/10.5840/soctheorpract200834322.

Tagesspiegel. 23.03.2023. "Volksentscheid Am Sonntag: Wissenschaftler hält Klimaneutrales Berlin Bis 2030 für Nicht Machbar." *Tagesspiegel* 2023 (23.03.2023). https://www.tagesspiegel.de/berlin/volksentscheid-am-sonntag-wissenschaftler-halt-klimaneutrales-bis-2030-fur-nicht-machbar-9547250.html.

Tank, Lukas. 2022. "Against the Budget View in Climate Ethics." *Critical Review of International Social and Political Philosophy*, 1–14. https://doi.org/10.1080/13698230.2022.2070833.

Tavoni, Massimo, and Richard S. j. Tol. 2010. "Counting Only the Hits? The Risk of Underestimating the Costs of Stringent Climate Policy." *Climatic Change* 100 (3–4): 769–78. https://doi.org/10.1007/s10584-010-9867-9.

Tavoni, Massimo, and Giovanni Valente. 2022. "Uncertainty in Integrated Assessment Modeling of Climate Change." *Perspectives on Science*, 1–37. https://doi.org/10.1162/posc_a_00417.

Theile, Merlinde. 24.06.2021. "Was wollen die Bürger sich zumuten?" *Die Zeit*, 24.06.2021.

Thompson, Erica L., and Leonard A. Smith. 2019. "Escape from Model-Land." *Economics* 13 (1). https://doi.org/10.5018/economics-ejournal.ja.2019-40.

Timperley, Jocelyn. 20.10.2021. "The Broken $100-Billion Promise of Climate Finance — and How to Fix It." *Nature News Feature*, 20.10.2021. https://www.nature.com/articles/d41586-021-02846-3.

Tol, Richard S. J. 2007. "Europe's Long-Term Climate Target: A Critical Evaluation." *Energy Policy* 35 (1): 424–32. https://doi.org/10.1016/j.enpol.2005.12.003.

Tollefson, Jeff. 2015. "Is the 2 °C World a Fantasy?" *Nature* 527 (7579): 436–38. https://doi.org/10.1038/527436a.

Trutnevyte, Evelina. 2016. "Does Cost Optimization Approximate the Real-World Energy Transition?" *Energy* 106: 182–93. https://doi.org/10.1016/j.energy.2016.03.038.

Trutnevyte, Evelina, Léon F. Hirt, Nico Bauer, Aleh Cherp, Adam Hawkes, Oreane Y. Edelenbosch, Simona Pedde, and Detlef P. van Vuuren. 2019. "Societal Transformations in Models for Energy and Climate Policy: The Ambitious Next Step." *One Earth* 1 (4): 423–33. https://doi.org/10.1016/j.oneear.2019.12.002.

UNEP. 2022. *The Closing Window: Climate Crisis Calls for Rapid Transformation of Societies*. Vol. 2022. The Emissions Gap Report. Nairobi: UNEP; United Nations Environment Programme. https://wedocs.unep.org/bitstream/handle/20.500.11822/40874/EGR2022.pdf?sequence=1&isAllowed=y.

UNFCCC. 1992. "United Nations Framework Convention on Climate Change." http://unfccc.int/resource/docs/convkp/conveng.pdf.

United Nations General Assembly. 2015. "The Paris Agreement." http://unfccc.int/resource/docs/2015/cop21/eng/l09r01.pdf.

Valentini, Laura. 2012. "Ideal Vs. Non-Ideal Theory: A Conceptual Map." *Philosophy Compass* 7 (9): 654–64. https://doi.org/10.1111/j.1747-9991.2012.00500.x.

van Beek, Lisette, Maarten Hajer, Peter Pelzer, Detlef van Vuuren, and Christophe Cassen. 2020. "Anticipating Futures Through Models: The Rise of Integrated Assessment Modelling in the Climate Science-Policy Interface Since 1970." *Global Environmental Change* 65: 102191. https://doi.org/10.1016/j.gloenvcha.2020.102191.

van Beek, Lisette, Jeroen Oomen, Maarten Hajer, Peter Pelzer, and Detlef van Vuuren. 2022. "Navigating the Political: An Analysis of Political Calibration of Integrated Assessment Modelling in Light of the 1.5 °c Goal." *Environmental Science & Policy* 133: 193–202. https://doi.org/10.1016/j.envsci.2022.03.024.

van de Ven, Dirk-Jan, Shivika Mittal, Ajay Gambhir, Robin D. Lamboll, Haris Doukas, Sara Giarola, Adam Hawkes, et al. 2023. "A Multimodel Analysis of Post-Glasgow Climate Targets and Feasibility Challenges." *Nature Climate Change* 13 (6): 570–78. https://doi.org/10.1038/s41558-023-01661-0.

van der Sluijs, Jeroen P., Arthur C. Petersen, Peter H. M. Janssen, James S. Risbey, and Jerome R. Ravetz. 2008. "Exploring the Quality of Evidence for Complex and Contested Policy Decisions." *Environmental Research Letters* 3 (2): 024008. https://doi.org/10.1088/1748-9326/3/2/024008.

van Diemen, Renée, J. Robin B. Matthews, Vincent Möller, Jan S. Fuglestvedt, Valérie Masson-Delmotte, Carlos Méndez, Andy Reisinger, and Sergey Semenov. 2022. "Annex i: Glossary." In *IPCC, 2022: Climate Change 2022: Mitigation of Climate Change. Contribution of Working Group III to the Sixth Assessment Report of the Intergovernmental Panel on Climate Change.* Cambridge, UK; New York, NY, USA: Cambridge University Press.

van Laar, Eric de, and Jan Peil. 2009. "Positive Versus Normative Economics." In *Handbook of Economics and Ethics*, edited by Jan Peil and Irene van Staveren. Edward Elgar Publishing. https://doi.org/10.4337/9781848449305.00056.

van Sluisveld, Mariësse A. E., J. H. M. Harmsen, Nico Bauer, David L. McCollum, Keywan Riahi, Massimo Tavoni, Detlef P. van Vuuren, Charlie Wilson, and Bob van der Zwaan. 2015. "Comparing Future Patterns of Energy System Change in 2 °c Scenarios with Historically Observed Rates of Change." *Global Environmental Change* 35: 436–49. https://doi.org/10.1016/j.gloenvcha.2015.09.019.

van Sluisveld, Mariësse A. E., Mathijs J. H. M. Harmsen, Detlef P. van Vuuren, Valentina Bosetti, Charlie Wilson, and Bob van der Zwaan. 2018. "Comparing Future Patterns of Energy System Change in 2 °c Scenarios to Expert Projections." *Global Environmental Change* 50: 201–11. https://doi.org/10.1016/j.gloenvcha.2018.03.009.

van Soest, Heleen L., Detlef P. van Vuuren, Jérôme Hilaire, Jan C. Minx, Mathijs J. H. M. Harmsen, Volker Krey, Alexander Popp, Keywan Riahi, and Gunnar Luderer. 2019. "Analysing Interactions Among Sustainable Development Goals with Integrated Assessment Models." *Global Transitions* 1: 210–25. https://doi.org/10.1016/j.glt.2019.10.004.

van Vuuren, Detlef P., Jae Edmonds, Mikiko Kainuma, Keywan Riahi, Allison Thomson, Kathy Hibbard, George C. Hurtt, et al. 2011. "The Representative Concentration Pathways: An Overview: Climatic Change, 109(1–2), 5–31." *Climatic Change* 109 (1–2): 5–31. https://doi.org/10.1007/S10584-011-0148-Z.

van Vuuren, Detlef P., Michel G. J. den Elzen, Paul L. Lucas, Bas Eickhout, Bart J. Strengers, Baş van Ruijven, Steven Wonink, and Roy van Houdt. 2007. "Stabilizing Greenhouse Gas Concentrations at Low Levels: An Assessment of Reduction Strategies and Costs." *Climatic Change* 81 (2): 119–59. https://doi.org/10.1007/s10584-006-9172-9.

van Vuuren, Detlef P., Andries F. Hof, Mariësse A. E. van Sluisveld, and Keywan Riahi. 2017. "Open Discussion of Negative Emissions Is Urgently Needed." *Nature Energy* 2 (12): 902–4. https://doi.org/10.1038/s41560-017-0055-2.

Vanderheiden, Steve. 2008. *Atmospheric Justice: A Political Theory of Climate Change.* Oxford: Oxford Univ. Press.

Vaughan, Naomi E., and Clair Gough. 2016. "Expert Assessment Concludes Negative Emissions Scenarios May Not Deliver." *Environmental Research Letters* 11 (9): 095003. https://doi.org/10.1088/1748-9326/11/9/095003.

Väyrynen, Pekka. 2013. *The Lewd, the Rude, and the Nasty: A Study of Thick Concepts in Ethics.* Oxford Moral Theory. Oxford: Oxford Univ. Press.

Victor, David G., and Charles F. Kennel. 2014. "Climate Policy: Ditch the 2 °c Warming Goal." *Nature* 514 (7520): 30–31. https://doi.org/10.1038/514030a.

Viehoff, Daniel. 2014. "Democratic Equality and Political Authority." *Philosophy & Public Affairs* 42 (4): 337–75. https://doi.org/10.1111/papa.12036.

Viehoff, Daniel. 2017. "The Truth in Political Instrumentalism." *Proceedings of the Aristotelian Society* 117 (3): 273–95. https://doi.org/10.1093/arisoc/aox015.

Vinichenko, Vadim, Aleh Cherp, and Jessica Jewell. 2021. "Historical Precedents and Feasibility of Rapid Coal and Gas Decline Required for the 1.5°c Target." *One Earth* 4 (10): 1477–90. https://doi.org/10.1016/j.oneear.2021.09.012.

Vocabulary.com. 2023. "Feasible: Dictionary." https://www.vocabulary.com/dictionary/feasible.

Wack, Pierre. 1985. "Scenarios: Shooting the Rapids." https://hbr.org/1985/11/scenarios-shooting-the-rapids.

Weber, Max. 1904. "Die Objektivität Sozialwissenschaftlicher Und Sozialpolitischer Erkenntnis." *Archiv für Sozialwissenschaft Und Sozialpolitik* 19 (1): 22–87.

Weber, Max, Edwards A. Shils, and Henry A. Finch. 1949. *Max Weber on the Methodology of the Social Sciences*. Illinois: The Free Press of Glencoe.

Weitzman, Martin L. 2007. "A Review of the Stern Review on the Economics of Climate Change." *Journal of Economic Literature* 45 (3): 703–24.

Weitzman, Martin L. 2014. "Martin Weitzman on the Problem from Hell." https://www.youtube.com/watch?v=gp6MjtpLrhE.

Weyant, John. 2017. "Some Contributions of Integrated Assessment Models of Global Climate Change." *Review of Environmental Economics and Policy* 11 (1): 115–37. https://doi.org/10.1093/reep/rew018.

Weyant, John, Ocen Davidson, Dowlatabadi. Hadi, James E. Edmonds, Michael Grubb, Edward A. Parson, Richard G. Richels, et al. 1996. "Integrated Assessment of Climate Change: An Overview and Comparison of Approaches and Results." In *Climate Change 1995*, edited by James P. Bruce, Hoesung Lee, and Erik F. Haites. Cambridge: Cambridge University Press.

Wiens, David. 2015. "Political Ideals and the Feasibility Frontier." *Economics and Philosophy* 31 (3): 447–77. https://doi.org/10.1017/S0266267115000164.

Wilholt, Torsten. 2009. "Bias and Values in Scientific Research." *Studies in History and Philosophy of Science Part A* 40 (1): 92–101. https://doi.org/10.1016/j.shpsa.2008.12.005.

Williams, Bernard. 1985. *Ethics and the Limits of Philosophy*. Cambridge, Mass.: Harvard Univ. Press.

Wilson, Charlie, A. Grubler, N. Bauer, V. Krey, and K. Riahi. 2013. "Future Capacity Growth of Energy Technologies: Are Scenarios Consistent with Historical Evidence?" *Climatic Change* 118 (2): 381–95. https://doi.org/10.1007/s10584-012-0618-y.

Wilson, Charlie, A. Grubler, N. Bento, S. Healey, S. de Stercke, and C. Zimm. 2020. "Granular Technologies to Accelerate Decarbonization." *Science (New York, N.Y.)* 368 (6486): 36–39. https://doi.org/10.1126/science.aaz8060.

Wilson, Charlie, Céline Guivarch, Elmar Kriegler, Bas van Ruijven, Detlef P. van Vuuren, Volker Krey, Valeria Jana Schwanitz, and Erica L. Thompson. 2021. "Evaluating Process-Based Integrated Assessment Models of Climate Change Mitigation." *Climatic Change* 166 (1–2). https://doi.org/10.1007/s10584-021-03099-9.

Workman, Mark, Geoff Darch, Kate Dooley, Guy Lomax, James Maltby, and Hector Pollitt. 2021. "Climate Policy Decision Making in Contexts of Deep Uncertainty—from Optimisation to Robustness." *Environmental Science & Policy* 120: 127–37. https://doi.org/10.1016/j.envsci.2021.03.002.

Workman, Mark, Kate Dooley, Guy Lomax, James Maltby, and Geoff Darch. 2020. "Decision Making in Contexts of Deep Uncertainty—an Alternative Approach for Long-Term Climate Policy." *Environmental Science & Policy* 103: 77–84. https://doi.org/10.1016/j.envsci.2019.10.002.

World Mapper. 2019. "Unchanging Politics of Climate Change." https://worldmapper.org/unchanging-politics-of-climate-change/.

Young, Iris Marion. 2000. *Inclusion and Democracy*. Oxford Political Theory. Oxford: Oxford University Press.

Ypi, Lea. 2010. "On the Confusion Between Ideal and Non-Ideal in Recent Debates on Global Justice." *Political Studies* 58 (3): 536–55. https://doi.org/10.1111/j.1467-9248.2009.00794.x.

Zimm, Caroline, Thomas Schinko, and Shonali Pachauri. 2022. "Putting Multidimensional Inequalities in Human Wellbeing at the Centre of Transitions." *The Lancet. Planetary Health* 6 (8): e641–42. https://doi.org/10.1016/S2542-5196(22)00124-3.

Zimm, Caroline, Frank Sperling, and Sebastian Busch. 2018. "Identifying Sustainability and Knowledge Gaps in Socio-Economic Pathways Vis-à-Vis the Sustainable Development Goals." *Economies* 6 (2): 20. https://doi.org/10.3390/economies6020020.